FIRE OFFICER'S HANDBOOK OF TACTICS

STUDY GUIDE

FIRE OFFICER'S HANDBOOK OF TACTICS

STUDY GUIDE

JOHN NORMAN

Published by Fire Engineering Books & Videos
A Division of PennWell Publishing Company
Park 80 West, Plaza 2
Saddle Brook, NJ 07663
United States of America

Book design by Patricia Rasch
Cover design by Steve Hetzel

Printed in the United States of America

1 2 3 4 5 6 7 8 9 10

0-912212-87-X

Contents

Introduction

This study guide is intended to be used in conjunction with the textbook *Fire Officer's Handbook of Tactics, Second Edition* by John Norman. It is important to realize that the subject area it covers is vast, and that no fire officer will be able to learn everything there is to know about tactics just by reading one text and using its study guide. The officer candidate and officer alike must seek out all sources of knowledge and add them to the experience he gains. He must follow his department's policies and procedures, as well as any applicable codes and standards.

At the very least, however, this guide will enhance the lessons put forth in the text. They will be of benefit to an instructor in developing tests, as well as to students in reviewing the assigned material.

The study guide has been divided according to the text, with a total of eighteen chapters, plus a final exam. You will find questions, answers, and references to the pages in the text where the most pertinent information appears for each question.

PART I QUESTIONS

GENERAL FIREFIGHTING TACTICS

General Principles of Firefighting

QUESTIONS

1. What is the most basic principle of firefighting?

2. As a firefighter who needs to choose between rescuing a victim and extinguishing the fire, which of the following considerations will most greatly affect your decision?

   1. Are there enough personnel available to do both simultaneously?
   2. If there are not enough personnel available, can the victim be reached without requiring hoseline support?
   3. Would putting out the fire eliminate the need for rescue?
   4. Are there other units in quarters that can be called to perform the rescue?
   5. What is the fire loading?

   A. All of the above.
   B. 1, 2, and 3.
   C. 1, 2, 3, and 4.
   D. 3, 4, and 5.

3. List the following victims in the order that they should be removed, assuming a serious fire condition.

   1. Persons on the top floor.
   2. Persons on the fire floor, remote from fire.
   3. Persons on the floor above the fire.
   4. Persons in the immediate vicinity of the fire.
   5. Persons below the fire.
   6. Persons three floors directly above the fire, two stories below the top floor.

   A. 3, 4, 2, 6, 1, 5
   B. 1, 3, 4, 2, 6, 5
   C. 4, 3, 1, 6, 2, 5
   D. 4, 3, 1, 2, 6, 5

4. "The lives of the occupants of the immediate fire area must always be the highest priority when staffing is insufficient to accomplish rescue and removal simultaneously." This statement is:

   A. Correct, because the occupants of the fire area must always be the first rescued.
   B. Incorrect, because you should begin the easiest rescues first, then progress to more difficult ones.
   C. Incorrect, since rescues must begin where the greatest number of occupants are seen.
   D. Correct, if in fact the victims are savable.

5. Which elements of a coordinated fire attack will reduce or eliminate the life hazard?

   1. Removing all victims.
   2. Venting to draw fire away from victims.
   3. Confining the fire.
   4. Extinguishing the fire.

   A. 1 only.
   B. 1, 2, and 3.
   C. 1, 3, and 4.
   D. All of the above.

6. "Interior structure firefighting requires an aggressive attack at every incident." This statement is:

   A. Correct as written.
   B. Incorrect, since vacant buildings should be treated with caution.
   C. Incorrect, since aggressiveness should be avoided at all costs.
   D. Correct, except that firefighters should never enter a vacant building.

7. Which of the following describes the proper sequence of actions at most structure fires?

   A. Locate, protect exposures, rescue.
   B. Confine, extinguish, locate.
   C. Locate, confine, extinguish.
   D. Confine, rescue, extinguish.

8. At three o'clock in the morning, you arrive simultaneously with the first engine to find fire venting from two windows on the ground floor of a three-story brick and wood-joist warehouse. The fire is exposing a nearby "fireproof" cold-storage warehouse. Where should you place the first hoseline?

   A. Between the fire building and the exposure.
   B. Inside the exposed building.
   C. Inside the fire building.
   D. In front of the fire building supplying a master stream.

Size-Up

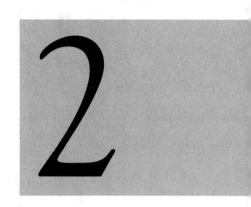

QUESTIONS

1. What is size-up?

2. Who performs size-up?

3. When should size-up be performed?

4. Given the following alert message, what element of the size-up can you ascertain? "Attention, Engines 1, 2, and 3; Ladder 1; and Chief 1. Respond to a house fire at 158½ Main Street. Fire reported on the third floor. The time is 1430 hours."

   A. There is no life hazard present.
   B. The fire is in a frame home.
   C. The fire is on the third floor.
   D. Three engines, one ladder, and one chief have been assigned.

5. In your community, on which side of the street would the even-numbered homes be found? Circle the correct choice.

   A. On the east side of a north-south street.
   B. On the west side of a north-south street.
   C. On the north side of an east-west street.
   D. On the south side of an east-west street.

6. A rule of thumb for anticipating collapse in ordinary construction or wood-frame buildings is that if fire has heavily involved an area for _____, fire forces should be withdrawn.

   A. Ten minutes.
   B. Fifteen minutes.
   C. Twenty minutes.
   D. Thirty minutes.

7. What are the main elements of the traditional thirteen-point or COAL WAS WEALTH size-up?

   1. C _____
   2. O _____
   3. A _____
   4. L _____
   5. W _____
   6. A _____
   7. S _____
   8. W _____
   9. E _____
  10. A _____
  11. L _____
  12. T _____
  13. H _____

8. A modernization of the traditional thirteen-point size-up would combine _____ and _____, and add _____ to the list.

9. What items from the size-up combine to produce the life hazard?

   1. Height.
   2. Weather.
   3. Time of day.
   4. Location and extent of fire.
   5. Occupancy.
   6. Apparatus and staffing.

   A. All of the above.
   B. 1, 3, 4, 6.
   C. 2, 3, 5, 6.
   D. 3, 4, 5.

10. What is the best method of reducing the life hazard before a fire?

11. Knowing the type of occupancy involved can tell an officer much about a given incident. Name five key variables strongly tied to occupancy.

12. As far as construction affects the size-up, a metal deck roof generally _____ the fire load.

   A. Eliminates.
   B. Reduces.
   C. Increases.
   D. Has little effect on.

13. A captain cited the following items as being major concerns about a building's construction that would affect the size-up. Which were correctly cited?

   1. The degree of compartmentation.
   2. The combustibility of the building.
   3. The number of hidden voids.
   4. The ability of the material to resist collapse.

   A. All of the above.
   B. 2, 3, and 4.
   C. 2 and 4.
   D. 4 only.

14. Rate the five classes of building construction in terms of their ability to withstand collapse.

15. Three factors may confuse your size-up of the height and area of the building. What are they?

16. All but one of the following choices represent special firefighting problems due to location. Which does not belong?

   A. Cellar fire in a fireproof building.
   B. A top-floor fire in a wood-frame apartment complex.
   C. A fire on the second floor of a three-story windowless warehouse.
   D. A fire on the second floor of a twelve-story high-rise hotel.

17. What three factors concerning smoke should be prime elements of a size-up?

18. List the minimum required fire flow for each occupancy below, expressed in gpm/100 square feet.

    1. Private home: ____ gpm/100 sq. ft.

    2. Stationery store: ____ gpm/100 sq. ft.

    3. Mattress factory: ____ gpm/100 sq. ft.

19. List four types of auxiliary appliances.

20. Complete the following descriptions of the effects that weather can have on firefighting.

    1. High heat and humidity rapidly _____.

    2. Below-freezing temperatures result in _____ and cause _____.

    3. High winds can _____ or force fire _____.

21. In the event that a department encounters a sewer trench in the fire block between the roadway and the fire building, the _____ should enter the block while the _____ remains out on the intersecting corner.

    A. Pumper, aerial device.

    B. Aerial device, pumper.

Engine Company Operations

QUESTIONS

1. Hose streams may be used for a variety of purposes, including which of the following?

    1. Fire extinguishment.
    2. Exposure protection.
    3. Controlled burning.
    4. Forced ventilation of gases.
    5. Absorbing toxic fumes.

    A. 1, 2, and 4.
    B. 1, 2, 3, and 4.
    C. 1 and 2.
    D. All of the above.

2. What are the three common methods of structure fire attack?

3. The preferred method of attack on a fire in the incipient stage is the _____ method.

4. Prior to beginning the actual attack on a free-burning fire, which actions should the nozzle team take?

    1. Attempt to locate and account for occupants.
    2. Survey the structure for alternate escape routes and other fire in remote areas.
    3. If you are approaching from below, take a quick look at the floor below the fire to get the layout.
    4. Begin ventilation above the fire.

    A. All of the above.
    B. 2, 3, and 4.
    C. 1, 2, and 3.
    D. 1 and 2.

5. At a working cellar fire in a 2½-story frame house, the first line should usually be stretched dry to which area before charging the hose-line?

   A. The front door.
   B. The top of the unenclosed cellar stairway.
   C. The bottom of the enclosed cellar stairway.
   D. The top of the enclosed cellar stairway.

6. While waiting for water, the nozzle team should look along the floor of the fire area for what?

   1. Possible victims.
   2. The location of the fire (glow).
   3. The layout of the area.
   4. The location of utilities.

   A. All of the above.
   B. 1, 2, and 3.
   C. 1 and 2.
   D. 1 only.

7. The method of attack best suited for the free-burning stage is the _____ method.

   A. Direct.
   B. Indirect.
   C. Combination.
   D. Alternative.

8. The indirect method of attack may be suitable for use on third-stage fires. All but one of the following are required ingredients to make this attack successful. Which one does not belong?

   A. High-heat condition present.
   B. Limited ventilation.
   C. Limited size of the potential fire area.
   D. Limited number of occupants.

9. Three strategic options in the firefighting plan include all but which choice?

   A. Offensive attack.
   B. Establish defensive positions.
   C. Take no action at all.
   D. Attack exposures.

10. What factors are primarily responsible for most fire spread to exposures?

   1. Radiation.
   2. Conduction.
   3. Convection.
   4. Direct flame contact.

   A. 1 only.
   B. 1, 2, and 3.
   C. 1 and 2.
   D. 1 and 4.

11. What strategy is required at a two-bedroom fire in an occupied apartment house?

12. An officer in command of the first attack line finds that he is unable to advance into the fire area due to heavy fire and tremendous heat. What three actions should he take?

13. The attack team referred to in Question 12 is still unable to advance the line after trying all of the options cited in your answer. What actions should the officer consider now?

14. Officers in command of operations expect to see some visual indication that the advancement of the hoseline is successful. All but one of the following should indicate that a line is successfully hitting fire. Which one does not belong?

   1. Flames darkening down.
   2. A change in the color of the smoke.
   3. A change in the volume of the smoke.
   4. A change in the movement of the smoke.
   5. The production of steam.
   6. The cessation of collapse.

   A. None of the above.
   B. 5 and 6.
   C. 3 and 6.
   D. 6 only.

15. The master stream of an elevating platform can cover the lower three floors of a structure for a length of how many feet (frontage) if positioned properly?

   A. 75 feet.
   B. 100 feet.
   C. 150 feet.
   D. Three times the length of the stream.

Hoseline Selection, Stretching, and Placement

QUESTIONS

1. What four factors affect the selection and placement of a hoseline?

2. What two factors determine whether a hose stream will be capable of extinguishing a given structure fire?

   1. The volume of the stream.
   2. The reach of the stream.
   3. The type of stream, whether fog or solid.
   4. The method of attack.

   A. 1 and 2.
   B. 1 and 3.
   C. 1 and 4.
   D. 3 and 4.

3. Compared with a smaller line, a larger hose-line exhibits all but one of the following characteristics. Which one does not belong?

   A. Greater flow potential.
   B. Greater reach and stream impact.
   C. Less flexibility.
   D. It puts out fire and cools well ahead of the members.

4. Residential occupancies have several characteristics that affect hoseline selection. Which of the following is not one of them?

   A. The need for speed.
   B. Low fire loading.
   C. The presence of dividing walls or partitions.
   D. The time of day.

5. All but one of the following areas present similar degrees of compartmentation and light fire loading. Which one does not belong?

   A. A "fireproof" hotel.
   B. A hospital patient care area.
   C. A high school classroom.
   D. A junior high storage area.

6. What are the factors that determine the diameter of the attack line?

    1. The occupancy of the building.
    2. The area of the building—i.e., the potential fire area.
    3. The size of the fire.
    4. The time of day.

    A. 1 and 2.
    B. 1, 2, and 3.
    C. 3 only.
    D. All of the above.

7. Which factors determine the required length of the attack line?

    1. The setback from the street.
    2. The height and area of the building.
    3. The presence or absence of an open stairwell.
    4. The presence or absence of a standpipe riser.
    5. The occupancy of the building.

    A. All of the above.
    B. 1, 2, 3, and 4.
    C. 1, 2, and 3.
    D. 1 and 2.

8. How much hose should you stretch for a fire on the top floor of a three-story factory building measuring 200 feet by 200 feet with a hydrant 100 feet from the front entrance?

    A. 500 feet.
    B. 600 feet.
    C. 900 feet.
    D. 1,000 feet.

9. Firefighters stretching to a fire on the fifth floor by way of an open stairwell in an old apartment house should ensure that at least how many lengths of hose are present inside the building? (Assume the stairway to be right at the front door.)

    A. Two.
    B. Four.
    C. Five.
    D. Six.

10. Stretching hose by rope on the outside of the building has all of the following advantages over stretching around a staircase except for which of the following?

    A. Speed.
    B. It requires fewer members to accomplish.
    C. It provides a ready means to stretch additional lines.
    D. It allows the members on the first line to attack from the fire escape and with the wind at their backs.

11. The degree of danger to a brick or concrete building from an exposure fire depends on:

    1. The number and size of the windows.
    2. The fire resistance of the walls.
    3. The proximity to the fire.
    4. The response of adequate numbers of fire department engine companies.

    A. All of the above.
    B. 1, 2, and 4.
    C. 2 and 4.
    D. 1 and 3.

12. What provides the greatest protection against radiant heat?

    A. Distance.
    B. Shielding.
    C. Light-colored or reflective paints.
    D. Water curtains.

13. Which isn't one of the main functions of a nozzle?

   A. To control the flow of the water.
   B. To divide the stream into fine droplets.
   C. To increase the velocity of the water.
   D. To change the shape of the water.

14. For maximum effectiveness, the discharge opening of a nozzle should not exceed what percentage of the supply diameter?

   A. 25 percent.
   B. 50 percent.
   C. 75 percent.
   D. 100 percent.

15. The disadvantages of using a cellar pipe on a severe cellar fire include all but which of the following?

   A. It requires constant supervision.
   B. The operation area may be untenable.
   C. The reach of its stream is limited, only fifteen to twenty feet.
   D. It only applies water in one or two directions at a time.

16. All but one of the following are prerequisites for high-expansion foam to darken down a fire. Which choice does not belong?

   A. The fire must involve Class A materials.
   B. The floor above the fire must remain tenable.
   C. A sufficient volume of foam must be applied.
   D. The foam must be able to reach the seat of the fire.

17. The bent applicator pipe is useful for discharging all but which agent?

   A. Water.
   B. Aqueous film-forming foam.
   C. Fluoroprotein foam.
   D. Dry chemical.

Water Supply

5

QUESTIONS

1. Refer to the diagram at the right. How would a gauge inserted at point A read compared with a similar gauge at point B?

   A. Lower.
   B. Higher.
   C. The same.
   D. The answer is indeterminable.

WATER

A

B

2. Refer to the diagram below. If a pumper were discharging into the hoseline below at a pump pressure of 150 psi, what would the pressure at gauge D read?

   A. 100 psi nozzle pressure.
   B. 125 psi.
   C. 150 psi.
   D. Higher than 150 psi.

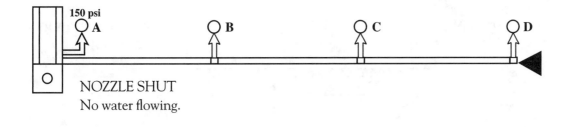

150 psi
A   B   C   D

NOZZLE SHUT
No water flowing.

3. Refer to the diagram at the right. What should the gauge at point X be reading?

   A. 2 psi.
   B. 8 psi.
   C. 12 psi.
   D. 18.4 psi.

4. Refer to the middle diagram. Gauge A through D all read the same. They illustrate what type of pressure?

   A. Elevation.
   B. Gravity.
   C. Head.
   D. All of the above.

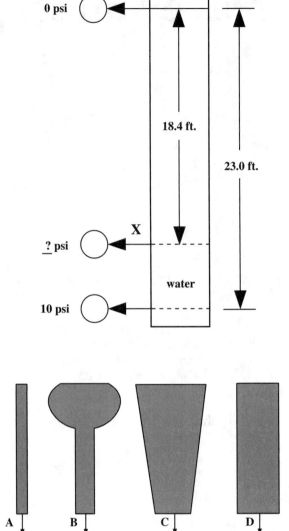

5. Refer to the diagram below. With water flowing at hydrant A, what should the residual pressure read at hydrant B?

   A. 45 psi.
   B. 55 psi.
   C. 70 psi.
   D. 95 psi.

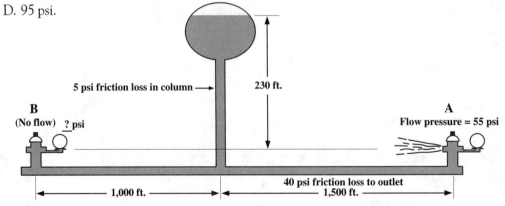

6. At high flows, most of what is commonly called "friction loss" is due to which of the following?

    A. Friction between the water and the hose lining.
    B. Turbulence within the water flow.
    C. Turbulence at the hose couplings.
    D. Gently curving bends in the hose.

7. Which of the following factors doesn't affect the water supply capability of a given layout?

    A. The capacity of the pump.
    B. The capacity of the hose.
    C. The available staffing.
    D. The capacity of the water source.

8. The minimum incoming pressure that a fire pump should receive is:

    A. 0 psi.
    B. 5 psi.
    C. 10 psi.
    D. 25 psi.

9. Where long distances or large flows are required of a relay operation, which choice doesn't solve the water delivery problem?

    A. Laying multiple supply lines.
    B. Changing the size of the discharge nozzle.
    C. Using large-diameter supply lines.
    D. Using a larger capacity pumper at the source.

10. What are the advantages of a telescoping platform over an aerial ladder when using an elevated master stream?

    1. Greater tip loading capacity.
    2. Greater discharge capacity.
    3. Higher elevation.
    4. Greater range of motion of the stream.
    5. The ability to reach over obstructions in front of a fire building.

    A. All of the above.
    B. 1, 2, 3, and 4.
    C. 1, 2, and 4.
    D. 1 and 2.

11. By dividing the flow of water from one line into two similar-size lines, the friction loss is reduced to what percentage of the loss for a single line?

    A. 10 to 15 percent.
    B. 25 to 35 percent.
    C. 45 to 55 percent.
    D. 75 to 85 percent.

12. One cubic foot of water weighs approximately:

    A. 42 lbs.
    B. 52 lbs.
    C. 62 lbs.
    D. 72 lbs.

13. The elevation pressure loss when pumping to an upper floor can be estimated at how many psi per floor, given 10 feet per story?

    A. 2 psi.
    B. 5 psi.
    C. 7 psi.
    D. 11 psi.

14. The maximum height that it is theoretically possible to draft water using a centrifugal fire pump is approximately:

    A. 25 feet.
    B. 20 feet.
    C. 33 feet.
    D. 35 feet.

15. What effect would tripling the length of the hoseline have on the total friction loss if the flow remained the same?

    A. None.
    B. Tripled.
    C. Nine times the loss.
    D. One-third the loss.

16. What effect does increasing the diameter of a hoseline have on the flow if the friction loss per foot remains constant?

    A. None.
    B. Doubled flow.
    C. Reduced flow.
    D. Increased flow.

Standpipe and Sprinkler Operations

QUESTIONS

1. According to NFPA statistics, sprinklers have successfully controlled or extinguished approximately what percentage of the fires at which they are present?

   A. 50 percent.
   B. 75 percent.
   C. 95 percent.
   D. 100 percent.

2. In more than ninety percent of all sprinkler operations, the fire is successfully contained by how many sprinklers?

   A. One or two.
   B. Five or six.
   C. Ten to twelve.
   D. Twenty to twenty-five.

3. How many cases of multiple deaths due to fire have occurred in buildings where operational wet-pipe sprinklers are present throughout?

   A. None.
   B. Dozens.
   C. Hundreds.
   D. Thousands.

4. Give two reasons for allowing the sprinklers to continue to operate until the fire has been extinguished.

5. A typical sprinkler head operating at 50 psi will flow approximately how many gpm?

   A. 20 gpm.
   B. 30 gpm.
   C. 40 gpm.
   D. 50 gpm.

6. The same sprinkler operating at 10 psi will flow approximately how many gpm?

   A. 12 gpm.
   B. 17 gpm.
   C. 22 gpm.
   D. 27 gpm.

7. What will be the effect if a pumper is drawing water from the same source as the sprinkler main?

8. What is the best way to prevent the drop in sprinkler flow?

9. What is the minimum supply that should be stretched to a siamese?

    A. One 2½-inch line.
    B. Two 2½-inch lines.
    C. One 3-inch line.
    D. One 5-inch line.

10. Which unit should supply the sprinkler siamese at a large working fire in a sprinklered building?

    A. The first-arriving engine.
    B. The second-arriving engine.
    C. An engine other than the one supplying hose streams.
    D. A multiple-alarm or mutual-aid engine.

11. What pressure should the pumper supply to lines feeding sprinkler siameses?

    A. 50 psi.
    B. 100 psi.
    C. 150 psi.
    D. 200 psi.

12. What pressure are new sprinkler systems normally tested to withstand?

    A. 100 psi.
    B. 150 psi.
    C. 200 psi.
    D. 250 psi.

13. All but one of the following are reasons why fires in sprinklered buildings are more smoky than fires in unsprinklered buildings. Which one does not belong?

    1. Unsprinklered fires are more often free-burning fires.
    2. Sprinklers don't allow complete combustion; thus, they create excess CO.
    3. Sprinklers cool the fire gases, making them less buoyant.
    4. Sprinkler spray patterns push down gases like a fog nozzle.
    5. Fires in unsprinklered buildings burn with less intensity.

    A. All of the above.
    B. 1, 2, 3, and 4.
    C. 2, 3, and 4.
    D. 2 and 4.

14. Which of the following isn't true of sprinklered fires?

    A. Increased production of carbon monoxide.
    B. No need for increased visibility.
    C. Sinking fire gases.
    D. Smoke being pushed down.

15. What unusual circumstance involving a sprinkler discharge is likely to cause a lot of grief for fire department personnel?

16. Where is the unusual circumstance alluded to in Question 15 likely to occur?

17. What effect do sprinklers located in skylights have on ventilation efforts?

18. List five steps necessary to operating effectively on a fire in a sprinklered building.

19. A system of overhead piping filled with water, connected to a water supply, and ready to discharge water at once through a heat-activated nozzle is what type of sprinkler system?

    A. Automatic wet pipe.
    B. Automatic dry pipe.
    C. Deluge.
    D. Preaction.

20. Where is the main drain valve located on a wet-pipe sprinkler system?

    A. Above the check valve.
    B. Below the check valve.

21. Where are automatic dry-pipe sprinklers usually found?

22. What are the primary differences between a dry-pipe valve and a wet-pipe alarm valve?

    1. The clapper of a dry-pipe valve is larger than that of a wet-pipe valve.
    2. The dry-valve clapper locks open when it's tripped.
    3. Dry valves must be reset manually.
    4. Dry valves contain only a small amount of priming water plus compressed air.

    A. All of the above.
    B. 1, 2, and 3.
    C. 2 and 3.
    D. 2 only.

23. What is a likely cause of a false alarm at a dry-pipe sprinkler system?

    A. Loss of air pressure.
    B. Loss of water pressure.
    C. Surge of air pressure.
    D. Water columning.

24. Both deluge systems and preaction systems are equipped with all but which of the following components?

    A. An OS&Y valve.
    B. A fire department connection.
    C. Open sprinkler heads.
    D. A valve actuated by detectors.

25. You would be most likely to encounter a preaction sprinkler system in which of the following occupancies?

    A. An aircraft hangar.
    B. The freezer of an ice cream factory.
    C. A computer room.
    D. A large nursing home.

26. What is the best method of determining whether a building is protected by a sprinkler system?

    A. The presence of a siamese connection.
    B. The sound of an alarm bell.
    C. Water discharge from drain pipes.
    D. Prefire inspection.

27. According to NFPA 14, standpipe systems may be of all of the types listed below except one. Which does not belong?

    A. Automatic wet standpipe.
    B. Automatic dry standpipe.
    C. Semiautomatic dry standpipe.
    D. Preaction standpipe.

28. All but one of the following sources of supply to sprinkler and standpipe systems are acceptable. Which choice does not belong?

    A. Suction tank, with or without pumps.
    B. Connection to water mains, with or without pumps.
    C. Gravity tanks.
    D. Pressure tanks.

29. At a serious fire on the upper floors of a stand-pipe-equipped building, what actions should you take to supply the standpipe system?

30. A standpipe system designed to provide first aid hose streams for occupant use, as well as to supply a 2½-inch fire stream, is a _____ system.

    A. Class I.
    B. Class II.
    C. Class III.
    D. Class IV.

31. Standpipes that provide hose for occupant use must protect the untrained user from excessive pressure. The pressure reducers limit maximum operating pressure in 1¾-inch hose to how many psi in pre-1993 buildings?

    A. 50 psi.
    B. 80 psi.
    C. 100 psi.
    D. 150 psi.

32. Define reflex time.

33. What items are necessary for conducting operations from a standpipe system?

34. A fire on what floor or below should always be approached by walking up the stairway, while floors above this level should be reached via the elevator?

    A. 6th floor.
    B. 7th floor.
    C. 8th floor.
    D. 9th floor.

35. According to NFPA 14, standpipe systems equipped with booster pumps must have pressure control devices to limit the maximum pressure. This pressure limiter provides a maximum of how many psi at the top-floor hose outlet?

    1. In pre-1993 buildings: _____
    2. In post-1993 buildings with 1½-inch outlets: _____
    3. In post-1993 buildings with 2½-inch outlets: _____

Ladder Company Operations

7

QUESTIONS

1. What are the seven primary tasks of ladder company personnel as outlined in the acronym LOVERS U?

2. What are the benefits of having an established operational plan?

   1. It formalizes thinking and makes others think in advance of what must be done.
   2. It assigns a degree of priority to each element.
   3. It lets all of the members, not just the OIC, know the plan.
   4. It establishes accountability for one's actions.

   A. All of the above.
   B. 1, 2, and 3.
   C. 1 and 3.
   D. 1, 2, and 4.

3. What are the three specific items that should be provided as part of any operational plan?

4. All but which of the following tools should be provided at every worksite at a structure fire?

   A. A forcible entry bar of some type (halligan tool).
   B. A flathead axe.
   C. A pickhead axe.
   D. A hook or pike pole of some type.

5. When developing an operational plan, what element is important for ensuring success?

6. If it is necessary to assign a member to operate alone on the fireground, how should he be paired up?

7. You staffing level dictates that one member of your ladder company crew may have to operate alone for part of an operation. What factors must you consider when weighing this responsibility?

8. What is the predominant fire problem nationally?

9. All but one of the following are duties of an interior search team at a private dwelling. Which one does not belong?

    A. Search of the fire area.
    B. Forcible entry.
    C. Expose hidden fire.
    D. Roof ventilation.

10. What does the acronym VES stand for?

11. At 3 P.M. on a weekday afternoon, which room of a home is most likely to contain a victim?

    A. Living room.
    B. Kitchen.
    C. Den/playroom.
    D. Bedroom.

12. How can tools and responsibilities be assigned in a volunteer or paid on-call department?

13. What are five factors that affect the selection and placement of portable ladders?

14. What is the major advantage of an extension ladder compared with a straight ladder?

15. Where should you place the tip of a portable ladder if you are planning to enter a window?

    A. Alongside the top of the window on the upwind side.
    B. Alongside the top of the window on the downwind side.
    C. Just at or below the windowsill.
    D. One or two rungs above the windowsill in the window.

16. A disoriented member moves into a room to seek an escape route after being cut off by fire. How could members who placed ladders outside the building have best indicated which window to move toward?

17. NFPA 1932 specifies that a heat-indicating label be applied to new ladders, indicating that the area of the ladder has been exposed to how many degrees Fahrenheit?

    A. 150°F.
    B. 200°F.
    C. 250°F.
    D. 300°F.

18. What is the designed loading capacity for new ground ladders, other than folding and attic ladders?

    A. 250 lbs.
    B. 500 lbs.
    C. 750 lbs.
    D. 1,000 lbs.

19. For firefighters to reach their objectives safely, what do aerial ladders and telescoping booms require?

20. How far should an aerial ladder be extended above a roofline?

21. Choose the least correct statement concerning the placement of an elevating platform basket:

    A. On a flat roof with no parapet, place the basket just on the roof.
    B. For a parapet of moderate height, place the basket with the top rail even with the top of the parapet.
    C. For parapets over six feet high, try to find another location, or use a ladder to ascend and descend.
    D. Telescoping platforms placed on the roof must be retracted to clear the roof edge.

22. Which is the incorrect answer regarding the scrub area of an aerial ladder or telescoping platform?

   A. The greater the working height, the greater the scrub area.
   B. The farther that the apparatus is from the building, the greater the scrub area.
   C. Positioning too close to the building will reduce the scrub area on the lower floors.
   D. The scrub area depends on the length of the device and the number of sections.

23. What is the most effective way to position a rear-mounted device so as to achieve the maximum scrub area?

   A. Parallel to the building.
   B. Nosed in toward the building.
   C. Backed in toward the building.
   D. Approximately fifteen degrees off parallel to the building.

24. When no specific situation demands that an aerial device take a particular position, where should you spot the apparatus?

25. What is overhaul?

26. Give an example of precontrol overhauling.

27. What are the five senses used for overhauling?

28. What is a good guide for evaluating whether to open a wall or ceiling that lacks any other evidence of fire?

29. At a serious fire in a cockloft area, where should the firefighter making the opening in the ceiling be located?

30. When pulling tin ceilings, where should the initial openings be made?

31. In which situation would hydraulic overhauling be inappropriate?

   A. At a fire in an obviously abandoned vacant structure.
   B. At a fire smoldering between a wooden lintel and a brick bearing wall above in an occupied factory.
   C. At a church fire where flames had seriously damaged the truss roof in a previous fire.
   D. At a bedroom fire that penetrated the ceiling and involved three bays.

32. What is salvage?

33. When you suspect that fire may be hiding in a partition, what action should you take before opening up that partition, if possible?

34. How does the hoseline crew contribute to salvage?

35. The control of utilities is a simple description given to what is often a complex operation. Describe the safety precautions that should be used.

Forcible Entry

8

QUESTIONS

1. What factors should be included in the forcible entry size-up?

    1. The time of day.
    2. The occupancy.
    3. The types of tools available.
    4. The direction that the door opens.
    5. The location of the fire, the victim, and the door to be used.
    6. The type of door, jamb, and locks encountered.

    A. All of the above.
    B. 2, 3, 4, 5, and 6.
    C. 3, 4, 5, and 6.
    D. 3, 5, and 6.

2. What factors help decide which method of forcible entry to use?

    1. The need for speed.
    2. The type of door and lock.
    3. The tools and staffing available.
    4. The degree of damage that will be done by a particular method.

    A. All of the above.
    B. 1 only.
    C. 2 and 3.
    D. 2, 3, and 4.

3. Which isn't one of the four rules of forcible entry?

    A. Try before you pry.
    B. Don't ignore the obvious.
    C. Use the door the occupants use.
    D. Always force the door with the least damage possible.

4. Key cylinders more than _____ inches from the door edge indicate a potentially serious forcible entry problem.

   A. One.
   B. Three.
   C. Six.
   D. Nine.

5. When using conventional forcible entry techniques, what factor determines the mechanics of the operation and how the tools are used?

6. When forcing an inward-opening door using conventional methods with a tool with a pronounced bevel on the fork end, how should you position the bevel?

7. When forcing an inward-opening door using the brute force method, what is the first way to attack the door?

   A. Use a sledgehammer to knock the door off its hinges.
   B. Use a sledgehammer to knock a panel out of the center of the door.
   C. Use a sledgehammer to breach a hole in the wall next to the lock.
   D. Use a sledgehammer to drive a halligan tool into the gap between the door and jamb.

8. When removing an inward-opening door from its hinges, which hinge should you attack first?

9. An outward-opening door recessed into a wall opening should be attacked using which end of a halligan tool?

10. The most efficient means of removing lock cylinders is with the:

    A. K tool.
    B. Slide hammer.
    C. Halligan tool.
    D. Lock puller.

11. After the key cylinder has been removed, what is the next step toward opening the lock?

12. A police lock is found only on what type of door?

13. Through-the-lock forcible entry is indicated by all but which style of lock?

    A. Police lock.
    B. Fox lock.
    C. Rim lock.
    D. Pivoting dead bolt.

14. The K tool is ineffective at pulling the cylinder of which type of lock?

    A. Police lock.
    B. Fox lock.
    C. Rim lock.
    D. Pivoting dead bolt.

15. When using a key tool to manipulate the mechanism of a fox lock, the square shaft must be rotated in which direction?

16. Through-the-lock forcible entry should not be used under what circumstances?

17. On encountering a multilock door, what is the first action that you should take?

18. How can you determine whether a multilock is engaged or not?

19. Security gates include all but which type?

    A. Manual.
    B. Mechanical.
    C. Electrical.
    D. Hydraulic.

20. What is usually the key to forcing roll-up security gates?

21. All of the following are useful tools in forcing case-hardened padlocks except:

    A. 30-inch bolt cutters.
    B. Duckbill lock breaker.
    C. Miner's pick.
    D. Halligan tool.

22. What tool is the most suited to removing padlocks located at or above shoulder height?

    A. Torch.
    B. Pick.
    C. Duckbill lock breaker.
    D. Power saw with an aluminum oxide blade.

23. When you encounter a padlock that is heavily shielded by a steel guard, what is the best way to attack it?

24. The American 2000 series gate lock may be removed using all but one of the following tools. Which one does not belong?

    A. Power saw with an aluminum oxide blade.
    B. Torch.
    C. Pipe wrench and cheater.
    D. Duckbill.

25. When encountering a padlocked security gate, it is usually better to cut the padlock than the gate itself. All but one of the following are correct reasons for this statement. Which one does not belong?

    A. Cutting the gate requires a special saw blade.
    B. Cutting the gate doesn't maximize access and ventilation.
    C. It can take longer to cut the gate than to cut one or two padlocks.
    D. Cutting the gate destroys it, which wouldn't be justified by a minor fire or emergency.

26. What conditions usually require that a roll-up security door be cut open?

    1. When heavy fire is present behind the door.
    2. When the speed of the application of water is critical.
    3. When it is an electrically operated door with the controls inside the fire area.

    A. All of the above.
    B. 1 and 2.
    C. 1 and 3.
    D. 1 only.

27. Which of the following is not an advantage of the three-cut method of cutting a roll-up security door?

    A. It isn't hampered by wind tabs.
    B. It can be used on doors recessed into a wall.
    C. It guarantees clearing a larger area for access and ventilation.
    D. It requires less cutting than the inverted V cut.

28. Series locks are often found on commercial establishments with many exits. They can only be opened by personnel using which method?

    A. Through the lock.
    B. Conventional forcible entry.
    C. Any key from any of the other locks in the building.
    D. All of the above.

Ventilation

9

QUESTIONS

1. What is ventilation?

2. What are the two reasons for venting?

    1. Venting for fire.
    2. Venting for life.
    3. Venting for smoke.
    4. Venting for property.

    A. 1 and 3.
    B. 1 and 2.
    C. 2 and 4.
    D. 3 and 4.

3. What is the difference between the two types of ventilation referred to in Question 2?

4. "Immediate ventilation performed to draw the fire away from the life hazard" is a description of what type of ventilation?

    A. Venting for fire.
    B. Venting for life.
    C. Venting for smoke.
    D. Venting for property.

5. Ventilation delayed until a hoseline is in place is a description of what type of ventilation?

    A. Venting for fire.
    B. Venting for life.
    C. Venting for smoke.
    D. Venting for property.

6. Where should ventilation to allow hoseline advance occur?

7. Horizontal ventilation for a life hazard must be accompanied by what?

8. Which isn't one of the factors that influence the decision to use either horizontal or vertical ventilation?

    A. Life hazard.
    B. The size and location of the fire.
    C. The construction of the building.
    D. The effects of weather, especially humidity.

9. Which choice isn't an advantage of horizontal ventilation over vertical ventilation?

   A. It's faster and easier to perform.
   B. It's less costly to repair.
   C. It's more effective at low-heat fires.
   D. It's more effective at fires with heavy smoke.

10. Which fires are most likely to benefit from roof cutting?

11. Mechanical ventilation is definitely called for in which of the situations listed below?

   1. Smoldering fire in a stuffed chair or sofa.
   2. Fire in a room that has been controlled by a sprinkler system.
   3. A low-heat fire below grade.
   4. A store fire in a one-story commercial structure in which flames are blowing out every window.

   A. All of the above.
   B. 1, 2, and 3.
   C. 1 and 3.
   D. 2 and 3.

12. Mechanical ventilation involves all but one of the following. Which choice does not belong?

   A. Directing a fog stream out a window after knocking down the fire.
   B. Venting a skylight or scuttle cover over the seat of the fire.
   C. Placing a portable fan.
   D. Turning the building's HVAC system to the exhaust mode.

13. What is a problem that is common to all operations using any mechanical ventilation technique?

14. Which choice is not a possible drawback to venting with a fog stream?

   A. Unnecessary water damage.
   B. Ice hazard in freezing temperatures.
   C. Drain on a limited water supply.
   D. Very demanding in terms of the number of personnel required.

15. Which choice does not properly reflect a drawback of using a fan in the negative-pressure mode?

   A. It requires sealing the opening around the fan; otherwise, churning results.
   B. Objects drawn into the exhaust screen block the flow of air.
   C. The location in the doorway creates access and safety problems.
   D. Drawing combustible gases through the fan could ignite the gas.

16. What factors must you take into consideration when deciding to use positive-pressure ventilation?

   1. The life hazard that may be affected by fire or venting.
   2. The extent of fire.
   3. The availability of hoseline.
   4. The degree of confinement possible.
   5. Debris that might be drawn or blown into the fan.
   6. The equipment available for the job, including the power supply.
   7. The wind.

   A. 1 only.
   B. 1, 2, and 3.
   C. 1, 3, and 4.
   D. All of the above.

17. What is the best way to prevent mushrooming in multistory structures?

18. When breaking a skylight over a staircase, what actions must you take?

19. Arrange the following types of roof sheathing in order of the most stable to the least stable.

    1. Tongue-and-groove boards.
    2. Furring strips.
    3. Older plywood.
    4. New fire-retardant plywood that has decayed.

    A. 1, 2, 3, 4.
    B. 1, 3, 2, 4.
    C. 3, 1, 2, 4.
    D. 3, 1, 4, 2.

20. What dangers do hard roof coverings pose to firefighters?

21. What factors should indicate the location of a ventilation opening on a peaked roof?

    1. The direction of the wind.
    2. The location of the fire as observed en route to the roof.
    3. Visible hot spots.
    4. Obstructions on the roof.

    A. 1 only.
    B. 1 and 2.
    C. 1, 2, and 3.
    D. All of the above.

22. Draw a quick cut.

23. What are the seven rules for cutting roofs?

24. What style of roof is this?

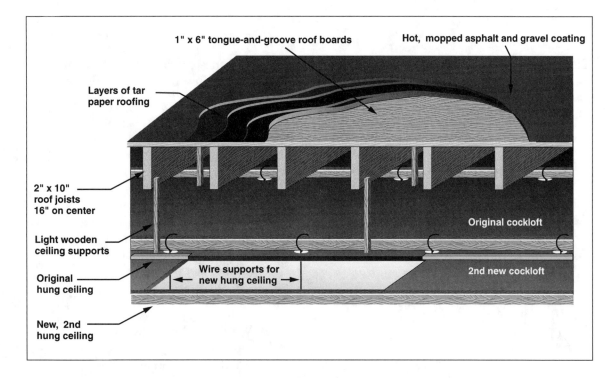

1" x 6" tongue-and-groove roof boards

Hot, mopped asphalt and gravel coating

Layers of tar paper roofing

2" x 10" roof joists 16" on center

Light wooden ceiling supports

Original hung ceiling

New, 2nd hung ceiling

Wire supports for new hung ceiling

Original cockloft

2nd new cockloft

25. What style of roof is this?

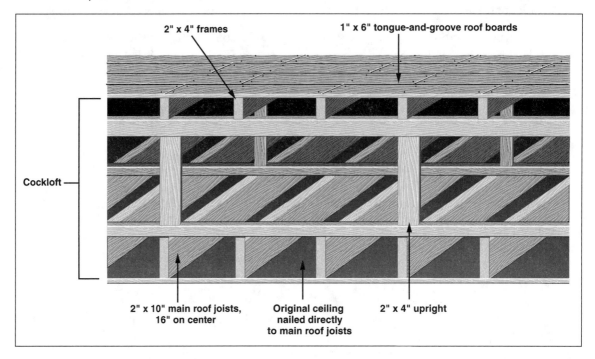

2" x 4" frames

1" x 6" tongue-and-groove roof boards

Cockloft

2" x 10" main roof joists,
16" on center

Original ceiling
nailed directly
to main roof joists

2" x 4" upright

26. What is the minimum commitment of personnel and equipment required to vent a working fire on the top floor of a large-area flat-roof structure?

27. For a serious top-floor fire in a flat-roof structure, what is the proper sequence of actions to be taken by the roof crew?

1. Vent over the stairs.
2. Cut the roof over the fire.
3. Vent top-floor windows.
4. Vent over other shafts.

A. 1, 2, 3, 4.
B. 1, 3, 2, 4.
C. 2, 1, 4, 3.
D. 1, 4, 3, 2.

28. What is a kerf cut?

29. What is the preferred style of examination hole to use in a roof to determine whether there is fire below?

A. Kerf cut.
B. Triangle cut.
C. Quick cut.
D. Basket cut.

30. All but one of the following correctly describe the effect of wind on determining where ventilation holes should be placed in a flat roof. Which does not belong?

A. Wind can blow fire toward the escape paths.
B. Wind can blow fire toward nearby exposures.
C. Wind can prevent extending a previously made vent hole.
D. Wind can prevent you from cutting the first holes in the best location if that location is upwind of subsequent cuts.

31. What is the proper sequence of cuts to complete the 4' × 4' vent hole shown below?

    A. C, F, A, E, D, G.
    B. C, A, B, D, E, G.
    C. D, E, A, G, B, C.
    D. D, B, C, A, E, G.

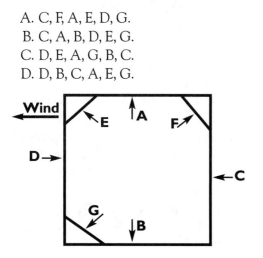

32. At a large-area flat-roof structure, such as a school, supermarket, or apartment building, what size ventilation opening is recommended for a serious fire?

    A. 4' × 4'.
    B. 6' × 6'.
    C. 8' × 8'.
    D. 12' × 12'.

33. What is the most likely part of a standard flat roof to fail as a result of fire?

    A. Roof boards.
    B. Roof joists.
    C. Roof covering.
    D. Roof trusses.

34. In an inverted roof, the roof boards are nailed to what element?

    A. A raised 2' × 4' framework.
    B. The top of the roof joists.
    C. The bottom of the roof joists.
    D. The top chord.

35. An inverted roof feels springy underfoot, even without a fire condition. If a serious fire has entered the cockloft of an inverted roof structure, is immediate evacuation of the roof necessary? Why?

36. What is a rain roof?

37. A rain roof poses all but one of the following dangers to firefighters during a fire. Which choice does not belong?

    A. The added weight may have overloaded supports, precipitating collapse.
    B. Rain roofs delay ventilation of the original cockloft area.
    C. The presence of two cocklofts slows fire travel.
    D. The presence of two cocklofts may result in conflicting estimates of a fire's intensity.

38. What is the most dangerous type of roof for a firefighter to work on or under?

    A. Standard flat roof.
    B. Inverted roof.
    C. Truss roof.
    D. Peaked roof.

39. Which of the following is not one of the major design flaws of all trusses, at least from a firefighting standpoint?

    A. The lack of mass compared with standard beams means earlier failure.
    B. The connectors often fail or speed the failure of the truss members.
    C. The failure of a single part does not destroy the entire truss.
    D. The open space within a truss allows rapid fire spread throughout the area.

40. What are the three elements of any truss?

41. What is the deadliest form of roof constuction per incident of all the current styles?

    A. Bar joist roof.
    B. 2 x 4 gusset plate trusses.
    C. Bowstring truss.
    D. Plywood I-beams.

42. What factors serve to identify bowstring trusses?

43. What is the biggest drawback of lightweight trusses?

44. What danger is posed to all flat-chord trusses by firefighting operations?

45. What problem is common to all open-web trusses, both wood and steel?

46. What factor makes plywood I-beams particularly dangerous to firefighters?

47. An incident commander arrives to find a serious fire in an occupied senior citizens development. He knows from preplanning that it is built using 2 x 4 trusses with gusset plates. He directs all of the following actions, but one of them is ill-advised. Which one?

    A. He orders the attack to begin from a distance, using the reach of the stream.
    B. He orders a line into the truss loft by opening the exterior wall.
    C. He orders total window ventilation and maximum lighting.
    D. He orders a rapid advance into the structure once the fire has been knocked down.

48. A trench cut should be used for what purpose?

49. All of the following elements of an attack, except one, are needed to make a trench cut successful. Which one does not belong?

    A. A large vent hole must be opened over the fire.
    B. The ceiling must be fully opened below the trench.
    C. Defensive hoselines must be in place on the top floor as well as on the roof.
    D. Additional vent holes must be cut on the safe side of the trench to ensure that fire has not gotten past the trench.

50. Making a trench cut consists of all but one of the following elements. Which one does not belong?

    A. Locating the trench.
    B. Cutting inspection holes.
    C. Cutting the trench.
    D. Cutting the main vent hole.

51. The newer-style thermopane windows pose all but one of the following dangers to firefighters. Which choice does not belong?

    A. The newer thermopane windows keep more heat in the building.
    B. They are more difficult to vent, especially with a hose stream.
    C. It is very difficult to remove the sash.
    D. They do not allow extension to the floor above.

52. What is one of the greatest dangers to firefighters working on venting a roof?

53. What is the most important action a member can take when operating a power saw?

54. What factor should be suspected by firefighters at a fire in a jewelry store or other high-value occupancy in a high-crime area?

Search and Rescue

10

QUESTIONS

1. Name the two phases of search.

2. What is the major difference between the two phases of search?

3. A quick search for live victims in the most likely areas, before the fire has been brought under control, describes which phase of search operations?

4. A search performed to ensure that there is absolutely no possibility that any victims remain in the structure is called a _____.

5. The primary search of the fire area is usually best begun where?

   A. At the door to the area, then working in toward the fire.
   B. By moving quickly toward the fire, then working back toward the door.
   C. Inside the fire room.
   D. Around the perimeter of the entire building.

6. Where should you begin searching when entering the area directly above the fire?

   A. At the door to the area, then working in toward the fire.
   B. By moving quickly toward the fire, then working back toward the door.
   C. Inside the fire room.
   D. Around the perimeter of the entire building.

7. To which areas should you give primary attention when searching residential structures?

   A. Bedrooms.
   B. Paths of egress.
   C. Stairways.
   D. All of the above.

8. What methods can you use to improve the conditions in the search area when performing a primary search?

   1. Close the doors between yourself and the fire.
   2. Vent windows in the area.
   3. Close the door to the fire room, if possible.

   A. All of the above.
   B. 1 only.
   C. 2 only.
   D. 1 and 3.

9. Describe how to recognize a child's crib in total darkness.

10. Describe a bunk bed.

11. As the incident commander, you find yourself confronting heavy fire on the ground floor of a three-story brick and wood-joist row house, with two apartments per floor. How many personnel should you assign to the search?

   A. Two.
   B. Six.
   C. Twelve.
   D. Twenty-four.

12. To perform a guide-rope search, what items should members have on hand before entering the search area?

   1. Two-way radios.
   2. SCBA, a PASS device, and a flashlight for each member.
   3. Search rope (200 feet or more of light line).
   4. Forcible entry tools.
   5. A large floodlight at the entrance.

   A. 1 and 2.
   B. 1, 2, and 4.
   C. 2 only.
   D. All of the above.

13. Is it necessary for all members to remain in contact with the guide rope?

14. Select the incorrect statement regarding team search operations.

   A. During team search, all members must be connected to the guide rope, either directly or by a short rope.
   B. The team must operate within set time limits.
   C. A control man remains outside to monitor the team.
   D. A rescue team must be standing by at the entrance to assist as needed.

15. What precaution is necessary when changing direction while using a search or guide rope?

16. What advantage does a thermal imaging camera have in fireground search?

17. Name three methods that may be helpful toward locating trapped or unconscious firefighters.

18. Which of the following is not a drawback of feedback-assisted rescue?

   1. The missing member must be radio-equipped.
   2. It takes time to implement.
   3. It requires an alternate means of communicating.
   4. It ties up what may be limited radio channels.
   5. It requires that members be recognized as being in trouble.

   A. None of the above are drawbacks.
   B. 3 and 4.
   C. 3 only.
   D. All of the above are drawbacks.

PART II QUESTIONS

SPECIFIC FIRE SITUATIONS

Firefighter Survival

11

QUESTIONS

1. All but one of the following are actions to be taken to reduce the firefighter mortality rate. Which does not belong?

   A. Improve hazard awareness and recognition training.
   B. Provide firefighters with emergency escape capability.
   C. Provide firefighters with additional breathing air.
   D. Provide firefighter rescue teams.

2. Which choice is not one of the three main rules of survival?

   A. Never get into a position in which you are depending on someone else to come and get you.
   B. Always know where your escape route is.
   C. Always know where your second escape route is.
   D. Always have a charged hoseline while searching.

3. While discussing the firefighters survival survey with his personnel, a company officer made the following comments about situational awareness. He was correct in which choices?

   1. You must be constantly evaluating the potential for flashover.
   2. You must be evaluating the potential for a backdraft.
   3. You must be aware of fire traveling in voids around you.
   4. You must be aware of the height of the ceiling above you.
   5. You must monitor the stability of the building.

   A. 1, 2, 3, and 5.
   B. 3, 4, and 5.
   C. 5 only.
   D. All of the above.

4. Members performing interior fire duty wearing SCBA must be able to free themselves from entrapments. What two maneuvers allow such an escape?

5. If you should find yourself entangled in cable TV wires while searching, which would not be a recommended action to take?

A. Inform your partners of the difficulty.
B. Try backing up, dropping lower, and proceeding forward.
C. Perform the emergency escape maneuver.
D. Remove your facepiece to untangle it from the wires.

6. Which choice would not be a reason for performing the emergency escape maneuver?

A. To free yourself from a serious entanglement.
B. To allow you to escape between the studs of a wall you breached.
C. To allow you to crawl out from beneath a beam blocking your exit.
D. To allow you, as a nozzleman, to enter an extremely narrow space while looking for fire.

7. If a firefighter finds himself cut off by rapidly spreading fire, he should take all but one of the following actions. Which does not belong?

A. Find an area of refuge and close doors between you and the fire.
B. Stay calm and stay put to conserve air.
C. Call for help by voice, radio, and PASS.
D. Locate any exit, even if you have to breach a wall or a window.

8. At times, a firefighter might find himself lost in a smoke-filled building where there is no danger of being overrun by fire but where a life-threatening atmosphere is present. This may be the case in a building that is sprinklered but where the sprinklers cannot fully extinguish the fire. Select the incorrect choice regarding the actions the lost firefighter should take.

A. He should stay calm and stay put to conserve air.
B. He should call for help by voice, radio, and PASS.
C. If he must seek an exit, he should walk with his light off so he can see other lights.
D. He should use all means at his disposal to attract attention to himself.

9. A firefighter who finds himself lost and running out of air must escape to fresh air very quickly. Four members in such a predicament each performed one of the following acts. Which members acted unwisely?

1. One member broke a window and dove headfirst down a ladder to avoid fire that was venting out of a window over his head.
2. One member kicked a door open with his feet to escape.
3. One member used a halligan tool to breach a wall to get into an adjoining apartment.
4. One member made an emergency bailout from a window by wrapping a personal rope around his chest.

A. All of the above.
B. 2 only.
C. 1, 2, and 4.
D. None of the above.

10. An officer instructing his members on the technique of the emergency bailout with a personal rope cited the following statements as being critical to success. This officer was incorrect in which statement?

    A. When descending, don't allow your hands to spread farther than about a shoulder-width apart.

    B. Remove all slack from the rope just prior to exiting.

    C. Never attempt this maneuver without a turnout coat, gloves, and a second member belaying the member descending.

    D. An NFPA-compliant one-person rope should be used for this escape.

11. One simple method of establishing accountability for the members of a career department is by using the written riding list. Select the incorrect statement regarding such a list.

    A. The list must be prepared at the start of each tour, and it must show all members working in the unit.

    B. The list must be updated to reflect any changes—e.g., if a member leaves sick.

    C. The list should be prepared in duplicate.

    D. The officer keeps his original for roll calls and gives the duplicate to the command post at large-scale operations.

12. What is the most important item on the fireground once a member has been reported missing?

    A. A riding list.

    B. Portable radios.

    C. A charged hoseline.

    D. A PASS device.

13. What are the prerequisites to a successful RIT operation?

    A. People, policies, tools, and techniques.

    B. People, planning, tools, and techniques.

    C. People, planning, and training.

    D. People, policies, and training.

14. According to the author, what is the minimum number of people needed to rescue a downed firefighter?

    A. Two.

    B. Three.

    C. Four.

    D. Six.

15. Which is the better choice of personnel for RIT duty?

    A. An engine company.

    B. A ladder company.

    C. An ambulance crew.

    D. A specially trained ladder crew.

16. Select a true statement regarding RITs.

    A. A RIT should automatically be dispatched on a unit's report of a working fire.

    B. A RIT should not be used for any duties until a firefighter is trapped.

    C. A RIT should always remain in the vicinity of the command post.

    D. A RIT should never be involved in using a handline.

17. Select an untrue statement regarding RITs.

    A. A RIT should be used as an attack unit if
       the incident escalates.
    B. A RIT should never be used for overhaul.
    C. A RIT that has been deployed for rescue
       duties must be replaced.
    D. A RIT must conduct a size-up and deter-
       mine the location of operating units.

18. All members of a RIT should step off their
    apparatus with certain equipment. Which of
    the choices listed below is incorrect?

    A. SCBA, plus an activated PASS device
       and a spare bottle.
    B. A large handlight, preferably on a sling for
       hands-free operation.
    C. A good, sharp knife.
    D. An NFPA-compliant personal rope.

19. A RIT officer should ensure that his six-
    person team reports in with which of the
    following tools?

    A. A 200-foot search guide rope for each
       member.
    B. A set of forcible entry tools.
    C. A spare mask for each member.
    D. A power saw, either wood- or metal-cut-
       ting, depending on the size-up.

20. At the scene, a RIT should attempt to locate
    all but which of the following items?

    A. A suitable ladder for the building.
    B. A copy of the building's floor plan.
    C. A copy of the command chart, indicating
       which units are operating and where.
    D. A hoseline that can be committed for
       RIT use, plus its supply.

21. As part of the RIT size-up, the team should
    have which of the following?

    1. A radio report while en route, giving the
       size, construction, and occupancy of the
       building, as well as the location of the
       fire.
    2. Someone monitoring tactical and com-
       mand channels for urgent messages and
       Mayday calls.
    3. On-scene reports of the building size-up;
       the occupancy; the location and extent of
       fire; the location of the units operating;
       and the routes of escape that those units
       may use.
    4. An idea of the progress of the operations.

    A. 1 and 2.
    B. 2 and 3.
    C. 1, 3, and 4.
    D. All of the above.

22. When a firefighter is trapped or missing, the
    incident commander must take some specific
    actions, all of which are correctly listed below,
    save one. Which of the following does not
    belong?

    A. Clear radio channels of any unnecessary
       traffic and establish clear lines of com-
       munication with the parties reporting the
       problem.
    B. Transfer command of either firefighting or
       rescue operations to another officer.
    C. Commit all needed resources to the res-
       cue effort and forego any firefighting.
    D. Special-call any other needed resources—
       e.g., ALS ambulances, additional ladder
       companies, etc.

23. As soon as the first report of a downed fire-fighter is received, all but one of the following actions should be taken. Which does not belong?

    A. A BLS ambulance should be called if one isn't on the scene.
    B. Additional firefighting personnel should be called.
    C. A protective hoseline and a spare mask should be brought to the vicinity of the operation.
    D. A resuscitator should be brought to the vicinity, even if there is no apparent need for it.

24. Several factors make the rescue of a trapped firefighter different from the rescue of a civilian. To which of the following does this rule not apply?

    A. All firefighters are full-grown adults.
    B. Turnout clothing and SCBA add weight and could get hung up on obstacles.
    C. There is added physiological stress involved in rescuing a firefighter.
    D. The trapped firefighter can at least help in his own rescue.

25. A rescue company officer conducting a drill on trapped firefighter rescue made the following statements. In which of these statements was he most correct?

    1. If fire conditions and logistics permit, use positive-pressure fans to provide clean air to the trapped members and rescuers.
    2. Begin multiple avenues of approach to trapped members.
    3. A clear chain of command is needed at these incidents.

    A. All of the above.
    B. 1 and 2.
    C. 2 and 3.
    D. None of the above.

26. Members performing RIT duty must be properly trained and equipped. Which statement below is incorrect regarding this duty?

    A. If you are the first to discover a trapped member, call for assistance immediately.
    B. Never share your mask with a trapped member.
    C. Be totally familiar with all types of SCBA used in your department.
    D. If you have dragged a victim out of immediate danger, begin basic life support and await aid.

27. When an unconscious firefighter must be removed from danger immediately, any of a variety of methods may be used. Select a true statement regarding such movement.

    A. The preferred drag is the feet-first method.
    B. If the SCBA keeps getting hung up on obstructions, take it off and place it on the member's chest.
    C. If the SCBA gets hung up, use the blanket drag.
    D. If the victim is entangled in an obstacle, the rescuer will have to feel all around the victim for the entanglement and may forcibly have to pull him free.

28. At a recent fire, an unconscious firefighter had to be brought down a flight of stairs to safety. Select the one correct statement below concerning this rescue.

    A. Whenever you have to drag an unconscious firefighter down the stairs, the blanket drag is the preferred method to use.
    B. One member starts down the stairs with the victim's head and shoulders, protecting these areas from further damage.
    C. The other member carries the legs and maintains balance for the group.
    D. Since gravity is helping under these circumstances, the smaller member should take the bottom position.

29. A RIT searching a smoke-filled second floor encounters a very large, unconscious firefighter who would have to be brought down a narrow, winding stairway. Which statement below is correct concerning the methods used to remove him?

   A. The RIT should await the arrival of the rescue company and their mechanical advantage system before continuing with this rescue.
   B. The rope should be attached to the victim's ankles.
   C. One rescuer drags the victim, feet first, down the stairs.
   D. The remaining members pull on the rope from below.

30. Choose the correct statement regarding the removal of a large, unconscious firefighter out of a window onto a portable ladder.

   A. It may be possible for one rescuer to crawl under the victim and use his thigh and back muscles to lift.
   B. If this rescuer can crawl under the victim, the only other help needed is one person on the ladder.
   C. If a rescuer cannot crawl under the victim, then the only choice is for three or four people to manhandle the victim up onto the windowsill.
   D. Ladder rescue can be made safer by using a stronger ladder.

31. At times it may be necessary to use a ladder as a high-point anchor so that members on the ground below can assist members on an upper floor to lift and haul a heavy person out of a window, particularly if the windowsill is high above the floor. Select the incorrect statement concerning this procedure.

   A. The rope is secured around the victim with a rescue knot and a slippery hitch around the chest.
   B. The rope leads out the window, over a rung of the ladder that is higher than the window, then down to the ground.
   C. On signal, three or four people pull from below while members in the room guide the victim to the windowsill, then the people on the ground lower the member to safety.
   D. This operation should not be performed unless the rope passes through a pulley on the ladder.

32. Select the most correct statement concerning the removal of an unconscious firefighter up a flight of stairs.

   A. The member is dragged to the base of the stairs.
   B. The member is secured to a ladder placed on the stairs.
   C. Several members at the top of the stairs grab a rope that is attached to the ladder.
   D. All of the members then pull the victim up with the ladder attached to him.

33. If extreme conditions exist, it might be necessary to haul up a person vertically through a hole in the floor. Choose the incorrect statement describing this maneuver.

    A. Another member must descend to the victim's location.
    B. The middle of a length of rope is lowered to a member below, who secures the victim with a handcuff knot around his wrists or ankles.
    C. If possible, lower a second rope, also by the middle, and place another handcuff knot on the other extremities.
    D. Once secured, on signal, four rescuers should haul the victim vertically through the opening.

34. Hauling a victim vertically by rope has several advantages. Select the choice below that is incorrect concerning this procedure.

    A. By hoisting on the extremities, the arms are always overhead, reducing the profile of the body.
    B. The shoulders are normally the widest part of the body.
    C. Hoisting by the arms anatomically serves to align the spine.
    D. If conditions permit, you should place a backboard at the mouth of the opening and secure the victim to it for transport.

Private Dwellings

QUESTIONS

1. Fires in one-and two-family homes account for what percentage of civilian fire deaths in the United States each year?

   A. 30 percent.
   B. 50 percent.
   C. 70 percent.
   D. 90 percent.

2. Place the following firefighting activities in the proper sequence.

   1. Secure the water supply.
   2. Perform VES of all areas.
   3. Stretch an attack line to protect the bedroom exits.
   4. Extinguish the fire.

   A. 1, 3, 4, 2.
   B. 3, 1, 4, 2.
   C. 1, 3, 2, 4.
   D. 3, 1, 2, 4.

3. What is the most serious defect, from a fire-spread standpoint, in a multistory private home?

   A. Open interior stairway.
   B. Combustible exterior walls.
   C. Large number of occupants.
   D. Small room size.

4. What is the second most serious defect in a multistory private home?

   A. Open interior stairway.
   B. Combustible interior finish.
   C. Large number of occupants.
   D. Small room size.

5. When planning water supply requirements for a well-involved two-story home without exposures, you should plan for about how many gpm?

   A. 100-200 gpm.
   B. 200-400 gpm.
   C. 400-600 gpm.
   D. 600-800 gpm.

6. The best tactic for securing a hydrant supply for a private-dwelling fire is _____ pumping, because of the _____ of the water _____.

   A. Positive, volume, delivered.
   B. In-line, volume, delivered.
   C. In-line, speed, delivery.
   D. Positive, speed, delivery.

7. When operating from a 500-gallon booster tank, how long can you maintain an effective fire stream through a 1¾-inch line?

   A. 1½ minutes.
   B. 2 minutes.
   C. 2½ minutes.
   D. 3 minutes.

8. Name three advantages of an in-line stretch.

9. A firefighter checking for extension over a serious basement fire gets to an outside wall, removes the baseboard molding, opens a hole in the wall, and discovers a 2 × 4 lying flat along the floor. What report should this member send to the incident commander regarding the building's construction?

   A. The wall is balloon-framed.
   B. The wall is brace-framed.
   C. The wall is platform-framed.
   D. None of the above.

10. Once fire is found spreading up a balloon-frame wall, speed is of the essence. Which of the following is a good action to take in this event?

    A. If high heat or fire is present, drive the hose stream up the bay.
    B. Position a hoseline on the top floor and one in the cellar.
    C. Quickly expose each bay for its entire height, especially those above and below the windows.
    D. Do not bother with roof ventilation if the fire is in the cellar.

11. All but one of the following are good reasons for calling for assistance to reinforce a ten-member crew at a house fire. Which does not belong?

    A. When fire is involving an attic or a cockloft.
    B. When fire has possession of two rooms and is blowing out the windows.
    C. When a firefighter is seriously injured.
    D. When a serious fire is present in a very large home.

12. An engine company driver, having laid an in-line supply from the most satisfactory hydrant, should spot his apparatus at which location?

    A. Just before the fire building.
    B. Directly in front of the fire building.
    C. Just past the fire building.
    D. At least two houses past the fire building.

13. Roof ventilation in a private dwelling would be justified in all but which of the following cases?

    A. Fire in the cellar of a balloon-frame house.
    B. Fire in the cellar of a platform-frame house.
    C. Fire that extends from the first floor via combustible siding to involve the eaves.
    D. Fire in an attic.

Multiple Dwellings

QUESTIONS

1. Multiple dwellings house at least how many families?

   A. Two.
   B. Three.
   C. Four.
   D. Six.

2. When do these occupancies present a high life hazard?

   A. Daytime.
   B. Nighttime.
   C. After midnight.
   D. All hours.

3. Multiple dwellings pose all of the following difficulties for firefighters except which one?

   A. Pipe recesses allow for fast vertical fire spread.
   B. The time delay required to cover large buildings.
   C. Recognizing converted dwellings.
   D. Combustible siding that allows rapid fire spread to cellars.

4. All of the following indicate the presence of a multiple dwelling except one. Which does not belong?

   A. A sign on the front saying Hotel Harriett Ann.
   B. The presence of four or five gas or electric meters.
   C. The presence of a doorbell and mailbox.
   D. Information received with the alarm, such as "Fire in apartment 3E."

5. What is an SRO?

6. All but one of the following building features promote fast-spreading fires in multiple dwellings. Which does not belong?

   A. Enclosed interior stairways.
   B. Light and air shafts.
   C. Unprotected shafts, such as dumbwaiters and compactors.
   D. Pipe chases and channel rails.

7. What are the two greatest threats to the structure in the event of fire?

  1. Pipe chases.
  2. Channel rails.
  3. Compactors.
  4. Dumbwaiters.

  A. 1 and 2.
  B. 1 and 3.
  C. 3 and 4.
  D. 2 and 3.

8. Why is the furred-out space around an I-beam more of a threat than a pipe chase?

  A. It is wider.
  B. It goes to the cockloft.
  C. It is harder to find.
  D. It is more likely to have fire enter it.

9. Immediately after knocking down a heavy fire condition in a ground-floor apartment in a four-story building, you note fire still burning in the wall behind the toilet. Which of the following would be the least correct action to take?

  A. Order the hose stream directed up and down the pipe chase.
  B. Contact other members on the floor above and tell them of the problem.
  C. Have the members on the roof feel the soil pipes for heat.
  D. If members on the roof report that one soil pipe is hot to the touch, order them to begin opening the roof around it.

10. Initial operations at multiple dwellings should focus on which of the following?

  A. Removing all of the occupants.
  B. Getting the first hoseline in place to protect the interior stairway.
  C. Raising portable ladders to all of the victims.
  D. All of the above.

11. You are the first to arrive at a fourth-floor fire in a six-story apartment house. The door to the fire apartment must be held closed while the occupants descend the open stairway. What action should you order your ladder company members to take while waiting for the remaining civilians to clear the stairway? (Note: The roof isn't open yet.)

  A. Vent the window on the landing of the stairway.
  B. Return to the street and raise ground ladders.
  C. Proceed to the floor above the fire for search.
  D. Return to the street and stretch another hoseline.

12. What is the best method of forcing entry at a multiple dwelling?

  A. Through the lock.
  B. Brute force.
  C. Conventional.
  D. With a hydraulic forcible entry tool.

13. What is the preferred size of attack hose for the average one-apartment fire in a Class IV multiple dwelling?

  A. 1½-inch.
  B. 1¾-inch or 2-inch.
  C. 2½-inch.
  D. 3-inch or larger.

14. When attempting to stretch a hoseline to an upper floor of a large building with several staircases, what action should you take to locate the proper stairway?

15. Of the three types of staircases, which is the greatest aid to firefighters?

   A. Isolated stairs.
   B. Wing stairs.
   C. Transverse stairs.

16. Of the three types of staircases, which is the greatest hindrance to firefighters?

   A. Isolated stairs.
   B. Wing stairs.
   C. Transverse stairs.

17. What type of stairway maintains a constant relationship to the floors and apartments on each level?

18. At a serious apartment fire, the second line might correctly be stretched to any of the following locations except one. Which choice does not belong?

   A. To the fire floor to back up the first line.
   B. To the fire floor, then to other involved apartments.
   C. To the fire floor by way of a route opposite the first line.
   D. To the floor above to handle extension.

19. At a serious top-floor fire, where the need for many additional handlines is evident, how should the third and fourth lines be stretched?

   A. By way of the interior stairway.
   B. By rope on the exterior of the building.
   C. From an outlet in a platform basket.
   D. Off the tip of an aerial ladder pipe.

20. Where heavy fire is evident in the cockloft of a large, multiwinged building, what defensive measures should you prepare?

   1. Position an elevating platform.
   2. Cut trenches to isolate the wings.
   3. Position additional vent holes to draw fire to the trench.

   A. All of the above.
   B. 1 and 2.
   C. 1 only.
   D. 2 and 3.

21. For a heavy cockloft fire, an elevating platform is a must. What are its main advantages?

22. At all serious fires in multiple dwellings, two members should immediately proceed to the roof. What is the correct sequence of actions that they should take?

   1. Force the bulkhead door.
   2. Vent any skylights atop the bulkhead.
   3. Search the stairway landing for victims trapped by a locked bulkhead.
   4. Examine the rear and sides for trapped occupants or extending fire.

   A. 1, 2, 3, 4.
   B. 1, 3, 2, 4.
   C. 2, 1, 3, 4.
   D. 4, 2, 1, 3.

23. What is the most preferable means of reaching the roof of a fire building located in the middle of a row of attached brick buildings?

   A. Aerial ladder.
   B. Fire escape.
   C. The interior stairway of the fire building.
   D. The stairway in an adjoining building.

24. For a serious lower-floor fire in a four-story apartment house, what actions should the roof team take after they have vented over all of the vertical shafts?

25. For a serious top-floor fire, what actions should the roof team perform after venting the vertical shafts?

26. What is a difficulty present in many renovated multiple dwellings?

27. What are three major collapse threats at multiple dwellings?

28. All but one of the following are difficulties present in Class I multiple dwellings. Which choice does not belong?

    A. Sturdier doors and locks make forcible entry more difficult.
    B. Early collapse of floors and roofs.
    C. The building holds tremendous heat, radiating it back at firefighters.
    D. Scissor stairs cause disorientation and hose-stretching problems.

29. What is the most likely method of fire extension in Class I multiple dwellings?

    A. Autoexposure through windows to the floor above.
    B. Flames burn through to the floor above.
    C. Flames burn through walls to the adjoining apartment.
    D. Flames travel via channel rails to the cockloft.

30. Class I buildings require extreme care when venting horizontally and a strong wind is blowing toward the fire apartment. All but one of the following are good actions to take toward evaluating the effect of venting the apartment windows. Which does not belong?

    A. If outside, check the direction of the wind.
    B. If in the fire area, vent upwind first, then downwind.
    C. Position all of the doors on the floor above the fire the same way that you intend them to be on the fire floor, open from the stairway and through to the fire area.
    D. If the wind will blow in, do not vent the windows.

31. In a Class I building, what action should you advise the occupants of apartments other than the fire apartment to take?

32. What is a method of controlling panic among building occupants who are hanging out of windows remote from the fire and calling for assistance?

33. As the incident commander, you arrive to find fire venting out of four windows (two apartments) on the third floor of a six-story Class IV multiple dwelling. What line should you order the first engine to stretch?

    A. 1¾-inch to the fire floor.
    B. 2½-inch to the fire floor.
    C. 2½-inch to the floor above.
    D. 1¾-inch to the floor above.

Taxpayers

14

QUESTIONS

1. The term *taxpayer* describes what type of structure?

2. What are the primary differences between old-style taxpayers and new-style strip malls?

   1. The amount of combustible materials present in the building.
   2. Strip malls are often noncombustible.
   3. Strip malls are often built without cellars.
   4. All taxpayers have two stories and a cellar.

   A. 1, 2, and 3.
   B. 1 and 2.
   C. 2 and 3.
   D. All of the above.

3. Which of the following is not one of the four styles of roof construction usually found on a taxpayer?

   A. Standard flat roof.
   B. Standard peaked roof.
   C. Metal deck on bar joist.
   D. Bowstring truss.

4. Why is a trench cut not recommended at a serious taxpayer fire?

5. A serious fire exists in a newer taxpayer built with a metal deck on a bar-joist roof. The roof team reports that the roof is sagging over the fire store. What orders should you give?

6. The presence of heavy fire should prompt the removal of all fire forces from the roof and interior of all but which of the following?

   A. A supermarket with a bowstring-truss roof.
   B. A gift shop built with lightweight trusses.
   C. A novelty store built with sawn floor and roof joists.
   D. A fruit store with plywood I-beams supporting the floor and roof.

7. Regardless of style, all taxpayers share all but one of these common dangers. Which does not belong?

    A. Difficult forcible entry.
    B. Potential backdrafts.
    C. Heavy fire load.
    D. Early collapse.

8. The rear of many taxpayers is heavily secured. What is often the fastest and least damaging means of gaining entry?

    A. Through the lock.
    B. Breaching the wall.
    C. Conventional forcible entry.
    D. Hydraulic forcible entry tool.

9. What is the proper sequence of operations to use at a taxpayer fire that shows indications of potential backdraft?

    1. Direct the stream onto the fire and advance slowly.
    2. Vent the storefront windows.
    3. Vent the cockloft and store from the roof before beginning the attack.
    4. Place charged 2½-inch hoselines out of the danger area of a blast.

    A. 3, 2, 4, 1.
    B. 3, 4, 2, 1.
    C. 4, 2, 1, 3.
    D. 4, 2, 3, 1.

10. All but one of the following structures are likely candidates for steel plating of the walls and roof. Which does not belong on the list?

    A. Jewelry store.
    B. Gun dealer.
    C. Electronics superstore.
    D. Furniture warehouse.

11. When fire is in the cockloft of a taxpayer, how many members and saws should be committed to the roof, at least initially?

    A. Two members, one saw.
    B. Two members, two saws.
    C. Four members, one saw.
    D. Six members, two saws.

12. Which of the following conditions would not result in a cellar fire in a taxpayer destroying the entire building?

    A. Heavy fire loading.
    B. Mazelike storage conditions.
    C. A lack of quick ventilation.
    D. The presence of a sprinkler system.

13. How much and what size hose should be ready at the front of a 75-foot-deep taxpayer before advancing into a cellar fire?

    A. 75 feet of 1¾-inch hoseline.
    B. 150 feet of 1¾-inch hoseline.
    C. 150 feet of 2½-inch hoseline.
    D. 200 feet of 2½-inch hoseline.

14. Describe the accuracy of the following statement, and explain your reasons: "When personnel are advancing the first line through the cellar and they encounter a high-heat condition, they should operate the line on wide fog to darken down the fire and to protect themselves."

15. What precautions are necessary to protect the members advancing a handline deep into the cellar of a taxpayer?

   1. A second line should be positioned in the cellar to protect their escape.
   2. A member should be positioned at the top of the cellar stairs to warn members of fire behind them or on the first floor.
   3. The line must always be charged before going down the stairway.

   A. 1 only.
   B. 2 only.
   C. 1 and 3.
   D. All of the above.

16. At a cellar fire in a taxpayer, the incident commander gave the following orders. He would be correct in which choices?

   1. Examine all voids and pipe chases for extension.
   2. Position a hoseline to cut off extension.
   3. Cut the first floor just inside the show windows for ventilation.
   4. Prepare distributors or cellar pipes in the event the handline attack fails.
   5. After a hole in the first floor has been completed, use a PPV fan to blow air down the cellar stairs.
   6. Set up a fog line on the first floor behind the vent hole to blow smoke, fire, and gases out the front window.

   A. 1, 2, and 3.
   B. 1, 2, 3, 4, and 6.
   C. 1, 2, 3, and 4.
   D. All of the above.

17. What is not one of the more serious threats to firefighters from a cellar fire in some taxpayers?

   A. Reinforced concrete floor.
   B. Unreinforced concrete floor.
   C. Terrazzo floor.
   D. Quarry-tile floor.

18. A six-inch-thick terrazzo floor that covers an area ten feet wide and ten feet long adds how much weight to the floor supports?

   A. 75 lbs.
   B. 1,500 lbs.
   C. 5,000 lbs.
   D. 7,500 lbs.

19. Which statement below is most correct concerning the tactics to use at a fire in a store or a strip mall?

   A. For very heavy fires, use 2½-inch handlines in the fire store and the exposed stores.
   B. A sufficient number of ten-foot hooks is needed to open ceilings rapidly.
   C. A line should be placed inside each exposed store.
   D. For heavy fire that has also entered the cockloft, an elevating platform should be used to apply a stream from above the roof.

20. Cockloft fires in taxpayers are extremely fast-spreading affairs that are compounded by a number of problems. Which of the following does not represent one of these problems?

   A. Multiple hung ceilings.
   B. Potential cockloft backdraft and ceiling collapse.
   C. Collapse of the parapet.
   D. The ease in locating the seat of the fire.

21. On entering a store, you encounter a heavy smoke condition at the ceiling with little or no visible fire and only moderate heat. Which of the following actions would be the least correct?

    A. Make a small examination hole in the ceiling just inside the entry to each area to see whether fire is over your head.

    B. Be sure to continue to poke upward until you are sure the members are hitting roof boards.

    C. After you pull the hook down, examine the head of the hook for signs of fire.

    D. If the hook shows evidence of fire overhead, immediately operate the line into the cockloft.

22. Why do parapets over display windows pose such a serious threat of collapse?

23. Besides their ability to topple parapets, what is another problem posed by expanding steel I-beams?

24. In commercial buildings, what advantages does a single large line have over two smaller lines?

25. If a serious fire has been burning for a prolonged time in a store or the cockloft, there is potential for collapse of the parapet. Which choice concerning this event is most correct?

    A. The collapse zone encompasses the entire sidewalk on all frontages.

    B. The parapets most at risk are on the two side walls.

    C. The heating of cast-iron columns precipitates most of these collapses.

    D. An elevating platform basket is not subject to this danger at a cockloft fire.

26. Which of the following are warning signs of potential backdraft?

    1. Heavy smoke issuing under pressure.
    2. Highly heated windows.
    3. Smoke puffing out and then being drawn back in.
    4. Large amounts of visible flame.

    A. All of the above.
    B. 1, 2, and 3.
    C. 1, 3, and 4.
    D. 1 and 3.

High-Rise Office Buildings

15

QUESTIONS

1. Some of the strategies designed for use in high-rise buildings may apply to many other structures as well. What are a few situations that might require high-rise tactics?

2. What items are included in the high-rise strategy?

   1. Determine the fire floor.
   2. Verify the location of the fire before committing handlines.
   3. Take control of the evacuation.
   4. Gain control of the building systems.
   5. Confine and extinguish the fire.

   A. 1, 2, and 3.
   B. 1, 3, and 4.
   C. 1, 3, 4, and 5.
   D. All of the above.

3. What factors make determining the fire floor and location so difficult in many high-rises?

   1. The central air-conditioning system that serves several floors.
   2. Open stairways.
   3. Smoke travel in elevator and utility shafts.
   4. Smoke travel in stairways due to the stack effect.

   A. 1 only.
   B. 1 and 2.
   C. 1, 2, and 3.
   D. All of the above.

4. The controlled evacuation of a high-rise building is a difficult task. Which of the following is not one of the main elements of this effort?

   A. Begin immediate evacuation of the fire floor and the floor directly above.
   B. Control the exit of those who are not endangered, particularly from the floors below the fire and out of the HVAC zone serving the fire floor.
   C. Search of the fire floor and all of the floors above the fire.
   D. Evacuation of the intervening floors (between the fire area and the top floors) is not necessary.

5. For effective operations, fire departments must be able to control a number of building systems at a high-rise fire. Which of the following isn't one of the major systems to be placed under direct fire department control during an incident?

   A. Elevators.
   B. HVAC systems.
   C. Electrical system.
   D. Communications.

6. Which isn't one of the distinguishing features of pre-World War II high-rises?

   A. Generally overbuilt, with structural members encased in concrete.
   B. Numerous exits usually remote from each other; often fire towers present.
   C. A lack of central air-conditioning and the presence of openable windows.
   D. Less compartmentation; no hung ceilings or blind spaces.

7. What was the most important feature of pre-World War II high-rises in limiting fire spread?

   A. Low fire loading.
   B. The absence of central air conditioners serving more than one floor.
   C. Floor-to-ceiling partitions of two-hour-rated construction.
   D. Concrete floor slabs on a metal deck and I-beams.

8. What effect does central air-conditioning have on fire operations in high-rises?

   1. The windows generally aren't readily openable for ventilation.
   2. Smoke, heat, and fire are readily spread from floor to floor, producing conflicting reports of the fire's location and sometimes panic among the occupants.
   3. The automatic smoke removal feature simplifies fire control.

   A. All of the above.
   B. 1 and 2.
   C. 2 only.
   D. 3 only.

9. What is the best way to prevent smoke travel through the HVAC system?

   A. Fire dampers inside the ducts.
   B. Smoke dampers inside the ducts.
   C. Smoke detectors in the intake ducts that automatically shut down the system.
   D. Manual shutdown of the system by building personnel on the report of a fire.

10. The HVAC system may be useful in smoke removal in high-rises. What must the incident commander know before using this system for that purpose?

11. If you arrive at a high-rise fire and receive reports of smoke on several floors, what action should you direct the building engineers to take regarding the HVAC system?

12. What two factors outweigh all others when performing ventilation during a high-rise fire?

   A. Wind and the stack effect.
   B. The height and area of the building.
   C. The fire condition and the stack effect.
   D. The wind and fire condition.

13. What is stack effect?

14. Is stack effect caused by a fire in a high-rise?

15. What effect does vertical ventilation have on the stack effect?

   A. Vertical ventilation will prevent mushrooming on upper floors.
   B. Vertical ventilation will draw fire and smoke away from shafts.
   C. Vertical ventilation allows smoke and heat to escape out of windows on the fire floor.
   D. Vertical ventilation forces smoke and heat back into the fire area.

16. Which of the following are reasons why elevator shafts should not be used for venting a high-rise building?

   1. It places an elevator out of service for fire department use.
   2. The open hoistway door would create a severe fall hazard.
   3. The hoistway door opening is too small to vent a serious fire effectively.

   A. All of the above.
   B. 1 and 2.
   C. 1 and 3.
   D. 2 and 3.

17. Two unusual problems are associated with high-rise ventilation. What are they?

18. Which of the following would not affect the decision regarding horizontal ventilation at a high-rise fire?

   A. The effect of wind.
   B. The stack effect, which can draw fire into the building instead of letting smoke and heat out.
   C. The status of the HVAC system.
   D. The effect of falling glass.

19. What is meant by the phrase "High, high; or low, low is a no-no"?

20. What is the major danger of using an elevator at a fire in a high-rise?

21. What is reflex time?

22. Which of the following choices is not a valid reason for using elevators during a high-rise fire?

   A. The need to reduce the reflex time.
   B. The safety and availability of a modern firemen's service elevator.
   C. The fatigue factor.
   D. The logistical problem of moving large numbers of people and their equipment up the stairs.

23. What two elements of elevator construction and design can make the use of elevators safer at a fire?

    1. The presence of sky lobbies below the fire floor.
    2. Firemen's service elevators.
    3. Blind-shaft elevators that go past the fire floor.
    4. The availability of freight elevators.

    A. 1 and 2.
    B. 1 and 3.
    C. 2 and 4.
    D. 1 and 4.

24. A chief officer lecturing on high-rise tactics to a group of company officers made the following statement: "The preferred hoseline for a large fire on an upper floor of a high-rise is the 2½-inch line equipped with a 1¼-inch solid tip." Justify this statement.

25. What is core construction, and why is it a problem for firefighters?

26. A chief officer arriving to find an office building fire with the entire 200' × 200' 30th floor fully involved in fire would know that his only chance of stopping the fire lies in what strategy?

    A. Pushing an aggressive, coordinated interior attack.
    B. Getting units onto the floor above to control extension until the fire burns itself out to a point that handlines can advance on the fire floor.
    C. Using outside streams from ground level.
    D. Using outside streams from the floor above.

27. What are two last-ditch methods of applying a stream to an upper floor of a high-rise fire when units are unable to enter the fire floor?

28. What difficulties are posed by scissor stairs in center-core construction?

29. What is the term for an open stairway connecting two or more floors within a single tenant's occupancy?

    A. Egress stair.
    B. Access stair.
    C. Return stair.
    D. Scissor stair.

30. What types of building features are recommended to safeguard an access stair?

31. What are the four major incident command designations at a high-rise fire?

32. Which of the following should be present at the operations post?

    1. Operations officer and an aide.
    2. Two separate radios, one on the frequency being used by attack forces, and the other in contact with the lobby command post.
    3. Copies of the floor plan for the fire floor and all the floors above.
    4. An additional means of communication with the command post, either by telephone or special radio.

    A. All of the above.
    B. 1, 2, and 4.
    C. 1, 2, and 3.
    D. 1 and 2.

33. What is a major function of the operations officer at a high-rise fire?

34. What are some of the requirements of a staging area at a high-rise fire?

35. When deciding where to establish a search and evacuation post, what would be a major concern?

36. What are the responsibilities of the search and evacuation officer?

37. What is a rule of thumb for determining the number of personnel required to staff a stairwell support section?

A. One member per floor up to the fire floor.
B. Two members per floor up to the fire floor.
C. One member per floor every two floors above the fire floor.
D. One member every two floors up to the fire floor.

Buildings Under Construction, Renovation, and Demolition

QUESTIONS

1. Buildings under construction, renovation, and demolition pose unique challenges to firefighters for all but which one of the following reasons?

   A. The size of the structure.
   B. The fire load.
   C. Structural problems.
   D. Firefighting operations.

2. All but one of the following are reasons why fires in buildings under construction, renovation, and demolition account for a large number of serious fires, as well as firefighter injuries. Which choice does not belong?

   A. Large amounts of concealed combustibles.
   B. Numerous sources of ignition.
   C. A lack of adequate fire protection features.
   D. An unlimited air supply if the windows are not intact.

3. Considering the potential dangers associated with construction, renovation, and demolition, which of the following explains why serious fires do not occur more often in such buildings?

   A. Most sources of ignition are only present when workers are gone for the day.
   B. Fires are spotted and extinguished in their incipient stage.
   C. Fire watches are posted around the entire site.
   D. The power to hazardous activities may be removed during work hours.

4. The preplanning of large construction sites is vital to fire suppression efforts. What are the key items to document when visiting such a structure?

    1. Large, undivided floor areas.

    2. Unenclosed vertical openings.

    3. Limited access to the upper floors.

    4. The storage of flammable gases in areas that are temporarily enclosed.

    5. Fire protection equipment is not keeping pace with construction.

    6. The building is partially occupied.

    A. All of the above.

    B. 1, 2, 3, and 6.

    C. 2, 3, and 5.

    D. 2, 3, 4, 5, and 6.

5. During severe fires, elevating platform streams can be extremely helpful if the fire is within their reach. What are some precautions that should be taken when using them?

    1. All personnel should be removed from all points where fire could be driven at them.

    2. All personnel should be removed from areas such as the streets below, where the master streams could propel large pieces of debris.

    3. Streams should be directed against scaffolding or shoring at close ranges so as to maximize their effectiveness.

    A. All of the above.

    B. 1 and 2.

    C. 1 and 3.

    D. 2 and 3.

6. It may be necessary to vary the application of master streams in buildings under construction or demolition as opposed to the pattern normally followed in ordinary buildings. Select the correct statement about this operation.

    A. Start the stream on lower floors and work up.

    B. Start the stream on upper floors and work down.

    C. Water damage is more of a concern in buildings under construction or demolition.

    D. Use wide fog patterns to maximize the production of steam.

7. Using hoists or temporary elevators at construction sites can cause serious problems. How can you tell an equipment hoist from a personnel elevator?

    A. Personnel elevators have guard rails.

    B. Personnel elevators have controls in the cars.

    C. Equipment hoists are automatic.

    D. Equipment hoists are larger.

8. Which is not one of the common difficulties found in the use of standpipe systems in buildings under construction or demolition?

    A. The siamese connections are blocked or hidden.

    B. The system is dry and all of the outlet valves are open.

    C. The top of the riser is uncapped, allowing water to flow right out the open top instead of through hoselines.

    D. The sectional valves are open below the fire, allowing water to flow right past them.

9. What is the danger of relying on membrane fire protection systems such as fire-rated hung ceilings in buildings under construction or renovation?

10. What is the problem with using spray-on fire-proofing on exposed steel structural members?

11. For how long after reinforced concrete is poured does the greatest danger of collapse exist?

   A. 12 hours.
   B. 24 hours.
   C. 48 hours.
   D. 28 days.

12. Select the most correct statement below concerning concrete construction.

   A. Poured concrete that is properly shored during the first 48 hours poses little danger in the event of fire.
   B. After 48 hours of curing, all of the shores may be removed from beneath the concrete.
   C. It takes approximately 28 days for concrete to cure fully.
   D. After concrete has fully cured, there is no danger from fire.

13. The practice of allowing tenants to occupy an area where there is work in progress on the floor below creates severe dangers. What is one method of safeguarding these occupants?

Fire-Related Emergencies: Incinerators, Oil Burners, and Gas Leaks

QUESTIONS

1. Natural gas is composed primarily of which of the following?

   A. Methane.
   B. Ethane.
   C. Propane.
   D. Nitrogen.

2. When attempting to determine whether natural gas is present or not, a sample can be analyzed in the laboratory. The presence of what two gases would indicate that the area contains natural gas?

3. Which statement about natural gas is incorrect?

   A. It is colorless.
   B. It is odorless.
   C. It is lighter than air.
   D. It is toxic.

4. Natural gas has a mercaptan compound added to it as an odorant. What problems can this create in the event of a very large leak aboveground?

5. What problem might this mercaptan additive pose in the event of an underground leak?

6. Which type of gas leak poses the greatest danger?

   A. A gas leak inside a structure.
   B. A gas leak outdoors.
   C. A gas leak with fire.
   D. None of the above.

7. What factors make the large transcontinental gas pipelines burn longer than the distribution main piping?

   1. Extremely high pressures (350-850 psi).
   2. Large-diameter pipes.
   3. Long distances between valves.
   4. Leaky valves.

   A. All of the above.
   B. 1 and 2.
   C. 1, 2, and 3.
   D. 2 and 3.

8. Low-pressure gas systems operate at what pressure?

   A. ¼ psi.
   B. 1 psi.
   C. 60 psi.
   D. 99 psi.

9. High-pressure gas systems use regulators to control the pressure on the appliance side of the regulator. If this device fails, what are the potential problems inside the affected structure?

10. The failure of a gas regulator can be recognized by the gas odor and a hissing sound coming from the vent. What actions should the fire department take?

    A. Notify the utility to stop the flow of gas by way of a remote control valve.
    B. Search the premises for fire, gas, and overcome victims, venting as required.
    C. Pull the electric meter to remove sources of ignition.
    D. If overcome victims are found while searching, immediately radio for assistance.

11. What danger did manufactured gas pose that natural gas does not?

12. What should be the fire department's first action on receiving a report of a gas leak?

13. When evaluating the danger of a reported gas leak, what two questions should you ask?

14. In the event that a gas leak is found in a stove, what is the preferred point of control?

    A. Appliance cock.
    B. Meter cock.
    C. Curb cock.
    D. Street valve.

15. An odorant is added to natural gas, allowing us to smell it about the time that it reaches what percentage of the lower explosive limit?

    A. 1 percent.
    B. 5 percent.
    C. 25 percent.
    D. 50 percent.

16. All but one of the following are potential sources of ignition for a mixture of flammable gas and air. Which choice does not belong?

    A. Keying a portable radio.
    B. The activation of a fire department pager.
    C. Turning off light switches.
    D. Static electricity generated by walking across a rug.

17. Under what circumstances might it be prudent not to vent immediately when gas is leaking into a structure?

18. Firefighters on the first engine to arrive at a serious indoor gas leak must be prepared to supply attack lines if an explosion occurs. Which is not a factor that affects the placement and use of these lines?

    A. The lines must be long enough to cover the entire building.
    B. The ability of the water supply to provide the needed flows.
    C. Apparatus should be positioned to protect members from the blast.
    D. Hoselines should be placed where they will immediately be able to enter the building.

19. Underground gas leaks can be especially dangerous to nearby structures due to what condition?

20. A leak in a plastic pipe is especially dangerous due to what phenomenon?

    A. Odorant filtration.
    B. High-pressure cracking.
    C. Freezing and thawing.
    D. Static electricity.

21. At an excavation site, a backhoe operator punctures a steel gas pipe, causing a leak that ignites. All of the following fire department actions would be correct at this scene except for which choice?

    A. Fire department members use dry chemical to extinguish the fire while rescuing the trapped backhoe operator.
    B. Fire department members use CO_2 to extinguish the fire so that utility workers can shut off a street valve.
    C. The fire department stands by until the gas company shuts down remote valves, stopping the flow.
    D. Fire department members under the protection of a water curtain attempt to plug the leak.

22. One gallon of liquid LPG when completely vaporized will form how many gallons of pure vapor?

    A. 27 gallons.
    B. 270 gallons.
    C. 1,700 gallons.
    D. 2,700 gallons.

23. As the temperature of propane liquid inside an LPG cylinder rises from 70°F to 100°F, what most nearly happens to the pressure exerted on the cylinder?

    A. The pressure rises 30 psi.
    B. The pressure rises 30 percent.
    C. The pressure doubles.
    D. The pressure quadruples.

24. What is the only way to prevent a BLEVE of a fire-exposed LPG cylinder?

25. An engine company approaching an LPG barbecue fire in the backyard of a private home should be prepared to supply how many and what size lines continually throughout the duration of the incident?

    A. One 1¾-inch.
    B. Two 1¾-inch.
    C. One 2½-inch.
    D. Three 1¾-inch.

26. In the event of a propane leak from a cylinder without fire, hose streams may be needed for what purposes?

27. How do oil burners get oil to burn very readily, even though the oil is well below its flash point?

28. When an emergency develops with an oil burner, how should the fire department stop its operation?

29. Units responding to find a smoky cellar with fire burning inside the oil burner should extinguish the flames by what method?

    A. Water fog.
    B. Foam.
    C. Dry chemical.
    D. None of the above.

30. What is the difference between a compactor and an incinerator?

Collapse

QUESTIONS

1. Rank the five classes of construction in descending order of resistance to collapse.

   1. Class I, fireproof.
   2. Class II, noncombustible.
   3. Class III, heavy timber.
   4. Class IV, ordinary construction.
   5. Class V, wood frame.

   A. 1, 2, 3, 4, 5.
   B. 1, 3, 4, 2, 5.
   C. 1, 3, 4, 5, 2.
   D. 1, 4, 3, 5, 2.

2. Generally, poured-concrete buildings are superior in their resistance to collapse. What is an exception to this statement?

3. Fires in heavy-timber structures generally do not endanger interior fire forces with structural collapse. What is a major exception to this statement?

4. What three concerns does steel pose when considering collapse potential?

5. What is a framed structure?

6. In which type of building, framed or unframed, is a collapse likely to be more serious?

7. What is a serious drawback of cast iron as a structural element?

8. Name at least six major causes of collapse at fires.

9. Name at least eight warning signs of potential collapse.

10. After the order to evacuate a structure has been given, what steps must the incident commander take to protect the members in the event collapse occurs?

11. Name one method used to warn all members to evacuate a structure immediately (emergency evacuation). What method does your department use?

12. How large should a collapse safety zone be?

13. What members must be monitored most closely to prevent them from encroaching on a collapse safety zone?

14. Where separation distances between buildings do not allow personnel to operate outside of the collapse zone, what other options are available for applying hose streams?

15. Which building requires the greatest collapse safety zone?

    A. A strip mall with a metal deck and bar-joist roof.
    B. A bowstring-truss automobile dealership.
    C. A supermarket with an inverted roof.
    D. A bowling alley with a bowstring truss roof and hip rafters resting on the end walls.

16. What are the five stages of the collapse rescue plan?

17. Before initiating any attempt to rescue victims in a collapse, the incident commander must perform a size-up of the scene. This is necessary to protect the victims and the rescuers, as well as to determine the best route to the victims. Name six items that must be considered during this size-up.

18. A victim tracking coordinator is essential at a collapse. What tasks does this person perform?

19. In which of the major types of collapse is it often simplest to locate a victim?

    A. Lean-to collapse.
    B. V-shaped collapse.
    C. Pancake collapse.
    D. Individual collapse.

20. In the event of an explosion that has pushed out one bearing wall, which type of collapse are you most likely to find?

    A. Lean-to collapse.
    B. V-shaped collapse.
    C. Pancake collapse.
    D. Individual collapse.

21. A collapse that results from beams burning through in the middle, or from an overloaded floor, would probably produce which type of collapse?

    A. Lean-to collapse.
    B. V-shaped collapse.
    C. Pancake collapse.
    D. Individual collapse.

22. Where are you most likely to find survivors in a lean-to collapse?

23. Selected debris removal, trenching, and tunneling are performed after all of the accessible voids have been searched. Why?

24. What is the main concern of the officer in charge of a group performing selected debris removal?

25. When attempting to tunnel through debris to reach trapped victims, multiple avenues of approach are recommended, if possible. What are some considerations that determine whether multiple tunnels are practical?

26. When a variety of tools that will all perform the same basic task are available at a collapse, what considerations will determine which tools to use?

27. What can be done if a wall or some other object is threatening to cause a secondary collapse?

28. Why must the general debris removal phase be performed under fire department supervision?

29. List six of the precautions to take while operating at a structural collapse.

PART III

FINAL EXAMINATION

Final Examination

QUESTIONS

1. Finding a heavy fire on the fifth floor of a six-story apartment building of ordinary construction, the officer in charge orders the members of the second-arriving engine to assist the first engine in stretching their hoseline, even though it's clear that a second line is required. The officer's actions are:

   A. Correct. Get the first line into operation as quickly as possible before stretching other lines.
   B. Incorrect, since the second line is needed and should be stretched immediately.
   C. Incorrect, since the members of the second engine should immediately begin evacuating the occupants.
   D. Correct, since two companies are needed to operate a hoseline under heavy fire conditions.

2. Among the reasons for selecting the combination method of attack, which of the following is of the greatest importance?

   A. It avoids the production of steam around the nozzle team.
   B. It disrupts the thermal balance tremendously.
   C. It puts firefighters in the best position to save lives.
   D. It exposes firefighters to the least danger.

3. With the new 1¾-inch or 2-inch hose and automatic nozzles, firefighters can maintain the same flow as with a 2½-inch line. This statement is:

   A. Correct as written.
   B. Correct, if the nozzle pressures are equal.
   C. Incorrect, if the nozzle pressures are equal.
   D. Incorrect, if a proper size nozzle is used on the 2½-inch line.

4. Of the following types of water pressure, which is of the most use to firefighters, given a hydrant-to-fire stretch?

   A. Static.
   B. Flowing.
   C. Residual.
   D. Head.

5. What is an unacceptable use for an elevating platform at a serious fire?

   A. Hauling a hoseline up to the fire on the fifth floor.
   B. Delivering power saws and personnel to the roof.
   C. Supplying a handline from a basket out-let on the floor above the fire.
   D. Using the master stream to knock down fire in the cockloft.

6. Positive pumping through large-diameter hose is the most efficient use of hose, pump, and aerial device. This statement is:

   A. Correct.
   B. Correct, if the pump discharge is the same as the nozzle rating.
   C. Correct, if the pump discharge rating is less than the nozzle rating.
   D. Incorrect.

7. What is the most common cause of false alarms involving wet-pipe sprinkler systems?

   A. Loss of air pressure.
   B. Loss of water pressure.
   C. Pressure surges in the supply.
   D. Vandalism.

8. Which of the following structural elements would not commonly be found in a new strip mall?

   A. Plasterboard walls.
   B. Concrete slab floors.
   C. Corrugated metal decks.
   D. Wooden roof joists.

9. A unit arriving at a large warehouse at 3 A.M. in response to a sprinkler valve alarm on a 0°F February morning notes the following: no alarm bell sounding, no water discharging from the drain piping, and no visible smoke or fire. They are advised by the alarm company that the valve has not reset. The most correct action would include which choices?

   1. Return to quarters immediately.
   2. Remain on the scene, awaiting the plant foreman.
   3. Force entry and begin a search of the premises.
   4. Send a knowledgeable radio-equipped member to locate the controls.
   5. Have the member at the controls imme-diately shut down the system that had been activated.

   A. 1 only.
   B. 2 only.
   C. 3 and 4.
   D. 3, 4, and 5.

10. What should be the primary source of water supply to a standpipe during fire department operations?

    A. Building fire pumps.
    B. Gravity tank.
    C. Pressure tank.
    D. Fire department pumper by way of a siamese.

11. A composite ladder is incorrectly described in which choice?

    A. Cannot conduct electricity.
    B. Does not tolerate high heat well.
    C. Has the same load-carrying capacity as aluminum.
    D. Only slightly heavier than aluminum.

12. Which of the following statements are incorrect regarding high-rise firefighting techniques?

    1. A separate pumper should be used to supply each siamese.

    2. Hoselines should be stretched to pump into lower-floor outlets if the siamese is damaged or if sufficient pressure cannot be developed through the siamese.

    3. The lower-floor outlets should be supplied at any serious fire as a precaution against total water loss in case the siamese or its piping fails.

    4. Fog nozzles operated off standpipe systems should be supplied by an engine pressure of 150 psi.

    A. None of the above.
    B. 2, 3, and 4.
    C. 3 and 4.
    D. 4 only.

13. What factors can cause smoke to travel counter to the normal diffusion pattern?

    1. Objects moving within shafts.
    2. HVAC system.
    3. Unusually warm weather.
    4. Fire in a very tall building.

    A. All of the above.
    B. 1 and 2.
    C. 1, 2, and 3.
    D. 2 only.

14. Under what circumstances should fire departments shut down sprinkler valves during a serious fire?

    1. When explosions have demolished sprinkler piping.

    2. When fire department handlines are in place to assume operations.

    3. When the fire has already overwhelmed a sprinkler system and the water supply is needed to protect exposures.

    4. When firefighters are unable to locate the fire due to the smoke and water spray.

    A. All of the above.
    B. 1, 2, and 3.
    C. 1 and 2.
    D. 1 and 3.

15. Which of the following are advantages of using a front-porch roof as a platform for VES?

    1. It provides a stable work platform.

    2. It provides a safe area of refuge if forced to retreat.

    3. Assistance is generally available in front of the building.

    4. Since the member is outside of the building, he won't need a mask.

    A. All of the above.
    B. 1, 2, and 3.
    C. 1 and 2.
    D. 1 only.

16. What is the most important item affecting ladder selection on the fireground?

    A. Nested length.
    B. Electrical conductivity.
    C. Strength.
    D. Reach.

17. During a fire in a high-rise building, who should be responsible for maintaining the proper pressure from the standpipe outlet?

    A. The first-arriving engine chauffeur.
    B. The officer in command of the hoseline.
    C. The officer in command of the fire.
    D. A radio-equipped member stationed at the outlet.

18. Choose the incorrect statement below concerning the proper positioning of aerial ladders and platforms for rescue when time is crucial.

    A. For a platform, stop the apparatus with the basket in line with the objective.
    B. Position the basket with the top rail just below the windowsill to allow easy access through the basket gate.
    C. Aerial ladders should be positioned with the turntable in line with the objective.
    D. The tip of the aerial ladder should be positioned just below the windowsill.

19. A chief officer lecturing several of his members about overhaul made the following statements. In which of them was he incorrect?

    1. Vertical voids, particularly those housing water, sewer, and vent pipes, as well as those for steel I-beams, must be the first priority when checking for extension.
    2. Once vertical extension has been checked, horizontal voids must be examined.
    3. Horizontal voids are usually found only on the top floor.
    4. At an old wooden home, be sure to open quickly the baseboards on the interior walls of the floor above the fire.

    A. All of the above.
    B. 2, 3, and 4.
    C. 3 and 4.
    D. 4 only.

20. Which of the following statements are true concerning the use of through-the-lock forcible entry?

    1. It is indicated by a heavy fire condition.
    2. It is indicated by specific types of doors.
    3. It is indicated by specific types of locks.

    A. All of the above.
    B. 1 and 3.
    C. 1 and 2.
    D. 2 and 3.

21. A hydraulic forcible entry tool should be used in lieu of conventional or through-the-lock methods in all but which of the following situations?

    A. When speed is essential.
    B. In poor visibility.
    C. On outward-opening doors.
    D. When damage is not a major concern.

22. All but one of the following indicate advantages of horizontal ventilation over vertical ventilation at most structure fires. Which choice does not belong?

    A. It's faster and easier than cutting the roof.
    B. It requires fewer personnel to accomplish.
    C. It provides superior smoke movement.
    D. It's less costly to repair.

23. At a serious fire on the ground floor of a four-story apartment house of ordinary construction, two members immediately ascend to the roof by way of an aerial device to vent over the staircase. After completing this operation and all other necessary ventilation of vertical arteries, these members should do which of the following?

    A. Descend the interior stairs and search the top floor.
    B. Feel the soil pipes that line up vertically with the fire area for heat.
    C. Reach over the parapet and vent all of the top-floor windows with a hook or pike pole.
    D. Immediately begin cutting the roof.

24. What is the most serious danger to a fire-fighter cutting a vent hole in a metal-deck bar-joist roof?

    A. Falling into the hole he's cutting.
    B. Being caught when large sections of the roof suddenly fail.
    C. Roof failure in under five minutes of fire exposure.
    D. Sudden failure of highly heated joists struck by hose streams.

25. A trench cut would be an effective defensive measure in which situation?

    A. A serious cockloft fire in a large super-market.
    B. A heavy fire blowing out one store in a row of twelve taxpayers.
    C. An H-type apartment building with a serious cellar fire.
    D. A long, narrow one-story warehouse with fire at one end.

26. Which of the following would not be part of the secondary search of a structure?

    A. Searching in bushes and shrubs outside a house.
    B. Checking inside a kitchen cabinet during overhaul.
    C. The nozzleman looking in along the floor while waiting for water.
    D. Checking inside each drawer in a dresser while performing salvage operations.

27. In which type of home would a cellar fire mandate a hoseline to the top floor or attic?

    A. Platform frame home.
    B. Braced frame home.
    C. Bubble-frame home.
    D. Balloon-frame home.

28. Your unit is ordered to search the floor above the fire in a large apartment house. As you ascend the interior stairway past the fire floor, you notice that the fire is definitely not yet under control. You also receive a radio report of persons trapped and jumping from the floor above. In this case, it would be best for you to do which of the following?

    A. Delay ascending the stairs until the fire is under control.
    B. Ascend immediately and force entry directly into the apartment directly over the fire apartment.
    C. Delay ascending until a charged hoseline has been brought to the floor above.
    D. Ascend immediately, but force entry into one or more apartments, other than the one directly over the fire, for use as areas of refuge if the hallway becomes impassable.

29. Commercial buildings share all but one of the following features with nonfireproof multiple dwellings. Which one does not belong?

    A. A common cockloft over the entire building.
    B. Large floor areas.
    C. Ordinary construction.
    D. A high life hazard regardless of the time of day.

30. Under which set of circumstances would it be proper to smash the front plate-glass windows of a row of commercial occupancies?

    A. When a light haze is present in the mid-dle store of a row of thirteen stores.
    B. When a store is puffing heavy smoke with no visible fire.
    C. In the third store in a row in which fire is visible, extending by way of the cockloft from two other fully involved stores.
    D. At a store at 3 A.M. where the windows are already blackened and cracked from the intense heat within.

31. A serious cellar fire in all but which of the following occupancies should cause the officer in charge to consider early collapse potential?

    A. An old-fashioned soda fountain with a terrazzo floor.
    B. A pizza parlor with an asphalt tile floor.
    C. A laundromat.
    D. A drugstore with a ceramic tile (mud) floor.

32. What creates the greatest headache for firefighters in postwar high-rise buildings?

    A. Difficulty in performing rapid effective ventilation.
    B. Lightweight construction techniques.
    C. Larger, wide-open floor spaces.
    D. Use of core-construction techniques.

33. During a serious high-rise fire, a crew of firefighters is sent to the roof. There they encounter the bulkheads of three separate staircases. They would be most correct if they were to avoid venting which stairway, at least for the time being?

    A. The evacuation stair.
    B. The ventilation stair.
    C. The attack stair.
    D. None of the above.

34. At a serious fire on the 33rd floor of a high-rise office building, a company officer made a number of statements regarding use of the elevators. Which is the most correct statement?

    A. Firemen's service elevators should always be used, since they are specially designed to protect firefighters.
    B. If there are no firemen's service elevators, use freight or service elevators for their larger load capacity.
    C. Each team that enters an elevator must have masks for each member, a portable radio, and a set of forcible entry tools.
    D. On entering an elevator car, a firefighter should immediately press the call-cancel button, then take the car directly to the floor below the fire.

35. Which of the following types of buildings poses the most severe danger of collapse in the event of a serious fire during the building's construction?

    A. Ordinary brick and wood-joist building.
    B. Precast concrete building.
    C. Steel-frame building with spray-on fireproofing.
    D. Poured-in-place reinforced-concrete building.

36. At a recent multiple-alarm fire involving a very large, high-pressure natural gas transmission main, firefighters took the following actions. They were correct in all but which choice?

    A. They evacuated the danger area.
    B. They used hose streams (solid tips at long distances) to protect exposed structures.
    C. They examined nearby structures for evidence of gas leaking into the building.
    D. They extinguished the fire only by closing a valve they located in a nearby street.

37. Which style of roof cutting is preferred for plywood roofs?

    A. The strip cut at the peak.

    B. The basket cut.

    C. The quick cut.

    D. The standard 4' × 4' cut.

38. Where should the first line be placed for a serious cellar fire in an occupied house that has vented out the cellar window and is now torching the asphalt siding on the rear wall, up toward the eaves?

    A. At the outside cellar doorway on the side of the house.

    B. At the rear cellar window to knock down the wall and operate into the cellar.

    C. Between the fire building and the most serious exposure.

    D. Through the front door to the interior cellar stairway.

39. While conducting the primary search of the floor above the fire in an SRO, you come across a number of doors that are padlocked on the hall side. You should:

    A. Force them all and search each room for occupants.

    B. Skip past them as if there is no life hazard within.

    C. Force them but move on—a later unit will search them.

    D. Search these areas only after obtaining the keys for each.

40. The danger of ceiling collapse in taxpayers is most prevalent in which of the following?

    A. Those built with lightweight truss roofs.

    B. Newer Class II (noncombustible) construction taxpayers.

    C. Newer Class IV (ordinary) construction taxpayers.

    D. Older taxpayers with multiple hanging ceilings.

41. A firefighter who finds himself lost in a large, smoke-filled warehouse in which the fire is being held in check by sprinklers should take all but one of the following actions. Which one does not belong?

    A. Move rapidly while searching for an exit.

    B. Call for help by voice, radio, and PASS.

    C. Use all means at his disposal to attract attention to his position.

    D. Lie with his face close to the floor and turn off his light for a few moments to look for other lights.

42. The Firefighters Survival Survey includes which of the following items?

    1. What is the occupancy?

    2. Where are the occupants?

    3. Where is the fire?

    4. How do we get in?

    5. How do we get out?

    6. What is happening to the building?

    7. What is the water supply?

    8. What is the status of auxiliary appliances?

    A. All of the above.

    B. 1, 3, 4, 6, and 7.

    C. 1, 2, 3, 4, 5, and 6.

    D. 1, 3, 7, and 8.

43. At minimum, when should a roll call be ordered?

    1. When a PASS alarm has been activated for more than thirty seconds.

    2. After the collapse of a member from an apparent heart attack.

    3. After sudden fire extension.

    4. After an emergency evacuation from a building.

    A. All of the above.

    B. 1, 2, and 3.

    C. 1, 3, and 4.

    D. 1 and 4.

44. Standpipes that provide hose for occupants to use must protect the untrained user from excessive pressure. The pressure reducers limit maximum operating pressure in 1¾-inch hose to how many psi in post-1993 buildings?

    A. 50.
    B. 80.
    C. 100.
    D. 150.

45. When encountering a heavy fire in a building with energy-efficient windows that have not yet been vented, when should horizontal ventilation not be performed?

    A. Immediately, while the attack team is in a safe area behind a closed door.
    B. After the attack hoseline has thoroughly cooled the area.
    C. Just before the nozzle team opens the nozzle.
    D. On reaching the windows from the inside, behind the nozzle.

46. Which of the following is not an advantage of positive-pressure ventilation?

    A. More efficient air movement.
    B. The fan doesn't clutter access to the building.
    C. PPV can safely remove flammable gases and vapors.
    D. PPV works best on advanced fires that have self-vented from many openings.

47. All but one of the following are reasons why bowstring trusses are extremely dangerous. Which does not belong?

    A. When a bowstring truss fails, it opens up a very large area almost instantly.
    B. In addition to its own failure, a collapsed bowstring truss will often produce a domino effect with other trusses.
    C. When the trusses fail, they often push out the end walls.
    D. They are difficult to detect from roof level.

48. Firefighters respond to a sprinkler alarm at a warehouse at 2 A.M. and are advised that the alarm has reset itself. Which type of sprinkler system is most likely in the building?

    A. Automatic dry pipe.
    B. Automatic wet pipe.
    C. Nonautomatic.
    D. Deluge.

49. Engine company personnel arriving at a working fire at a strip mall performed the following actions. Which of them might have acted more wisely?

    A. Engine 1 laid 200 feet of 5-inch hoseline and positioned directly across the street from the fire store.
    B. Engine 2 connected to a hydrant with its 6-inch soft suction and fed three handlines.
    C. Engine 3 laid in 400 feet of 3-inch hoseline to the rear alley to protect exposures.
    D. Engine 4 laid two lines of 3-inch hose from Tower Ladder 2 and connected to a hydrant around the corner.

50. A 2½-inch handline has several advantages over smaller lines when operating at a commercial building. Which of the following is not one of those advantages?

    A. The volume of water delivered and the reach of the stream.
    B. The ability to make tight bends and turns.
    C. The efficiency of personnel.
    D. The power of the stream to penetrate obstructions.

51. Which of the following are warning signs of potential backdraft?

    1. Heavy smoke issuing under pressure.
    2. Highly heated windows.
    3. Smoke puffing out, then being drawn back in.
    4. Large amounts of visible flame.

    A. All of the above.
    B. 1, 2, and 3.
    C. 1, 3, and 4.
    D. 1 and 3.

52. Despite all efforts to advance a handline into a cellar fire of an old downtown business, no progress is made after ten minutes due to the intensity of the heat. What orders should the incident commander give next?

    A. Deploy fresh personnel to renew efforts.
    B. Withdraw handlines from the cellar and place distributors or cellar pipes through holes cut in the wooden first floor.
    C. Introduce high-expansion foam through the cellar stairs.
    D. Flood the first floor with master streams.

53. Serious fires in strip malls can best be dealt with using elevating platforms. All but one of the following illustrate good use of an elevating platform. Which does not belong?

    A. Using a power saw from the basket to open the fascia of the overhang in front of the stores.
    B. Using the basket nozzle from the sidewalk to direct its stream into the cockloft.
    C. Members in the basket remove a sign on the front of an old taxpayer to expose the cockloft.
    D. Using the nozzle from above the roof to extinguish fire coming through holes in the roof.

54. What is the normal pressure in a gas stove fed by a high-pressure service?

    A. ¼ psi.
    B. 2½-60 psi.
    C. 99 psi.
    D. 350-800 psi.

55. Which of the following should the fire department not use for control if a serious gas leak occurs in a structure?

    A. Service cock.
    B. Meter cock.
    C. Curb cock.
    D. Street valve.

56. How far away from the end walls should the collapse safety zone extend at a bowling alley fire with 22-foot-high brick walls supporting a bowstring truss roof that has hip rafters resting on the end walls?

    A. At least 22 feet.
    B. 33 feet.
    C. 40 feet.
    D. 53 feet.

57. All but one of the following are prerequisites for high-expansion foam to darken down a cellar fire successfully. Which does not belong?

    A. The fire must involve only Class A materials.
    B. The floor above the fire must be vented.
    C. A sufficient volume of foam must be applied.
    D. The foam must be able to reach the seat of the fire.

58. What are the advantages of a telescoping platform over an aerial ladder for using an elevated master stream?

    1. Greater tip-loading capacity.
    2. Greater discharge capacity.
    3. Greater reach (height) of the tip.
    4. Greater range of motion of the stream.
    5. The ability to reach over obstructions in front of the building.

    A. All of the above.
    B. 1, 2, 3, and 4.
    C. 1, 2, and 4.
    D. 1 and 2.

59. Prior to beginning the actual attack on a free-burning fire, what actions should the nozzle team take?

    1. Quickly attempt to locate and account for occupants.
    2. Survey the structure for alternate escape routes and other fires in remote areas.
    3. If approaching from below, take a quick look in on the floor below to get the layout.
    4. Begin ventilation above the fire.

    A. All of the above.
    B. 2, 3, and 4.
    C. 1, 2, and 3.
    D. 1 and 2.

60. At three o'clock in the morning, you arrive simultaneously with the first engine to find fire venting from two windows on the ground floor of a three-story brick and wood-joist warehouse. The fire is exposing a nearby fireproof cold-storage warehouse. Where should the first line be placed?

    A. Between the fire building and the exposure.
    B. Inside the exposure.
    C. Inside the fire building.
    D. In front of the fire building, supplying a master stream.

61. A captain lecturing firefighters about collapse rescue on the fireground made the following statements. Which of them are correct?

    1. In the event that a member is trapped, all personnel should immediately come to assist in removing him.
    2. All firefighting and support functions should cease, and the resources be used to rescue the trapped member.
    3. All equipment is considered expendable when attempting to rescue a firefighter.
    4. Once all of the victims have been accounted for, shift back to a cautious approach, and make sure that everyone is informed.

    A. All of the above.
    B. 2, 3, and 4.
    C. 3 and 4.
    D. 4 only.

62. List the following selection of victims in the order in which they should be removed from the building. Assume that there is a serious fire on the second floor of a six-story 100' × 150' brick and wood-joist apartment house.

    1. Persons on the top floor.
    2. Persons on the fire floor, remote from the fire.
    3. Persons on the floor directly above the fire.
    4. Persons on the fire floor in proximity to the fire.
    5. Persons on the floor below the fire.
    6. Persons on the fourth floor.

    A. 4, 3, 2, 6, 1, 5.
    B. 4, 2, 3, 6, 1, 5.
    C. 4, 3, 1, 6, 2, 5.
    D. 4, 3, 1, 2, 5, 6.

63. A firefighter operating at an advanced fire in a wood-frame home watches as a door suddenly swings closed for no obvious reason. This should be taken as a warning sign for evacuation because of the possibility of:

   A. Backdraft.
   B. Flashover.
   C. Structural collapse.
   D. Poltergeist explosion.

64. All but one of the following situations pose special firefighting problems due to their location. Which does not belong?

   A. A cellar fire in a fireproof office building.
   B. A top-floor fire in a wood-frame apartment complex.
   C. A fire on the top floor of a windowless three-story warehouse.
   D. A second-floor fire in a four-story fireproof hotel.

65. A rule of thumb for anticipating the collapse of a building of wood-frame or ordinary construction is that if heavy fire has involved an area for _____ minutes, fire forces should be withdrawn.

   A. Ten minutes.
   B. Fifteen minutes.
   C. Twenty minutes.
   D. Thirty minutes.

66. Where long distances and large flows are required for a relay operation, which choice does not solve the water-delivery problem?

   A. Laying multiple supply lines.
   B. Using an automatic nozzle to use the available water most efficiently.
   C. Using large-diameter supply lines.
   D. Using a large-capacity pumper at the source.

67. Engine company personnel arriving first at an apartment house in response to a report of an explosion in the cellar on a bitterly cold night find the entire cellar knee-deep in a ghostly white mist that smells strongly of fuel oil. Which of the following should they not do?

   A. Evacuate the entire building.
   B. Immediately enter the cellar to shut down the oil at the tank valve.
   C. Stretch a handline with a fog nozzle and saturate the cellar.
   D. Remove all sources of ignition and vent the area with a PPV fan.

68. In the absence of a specified department protocol, a ladder company responds to an activated carbon monoxide detector in a home. Which of the following is correct?

   A. Question the occupants to determine whether anyone is feeling ill.
   B. Send a member with a CO detector to examine all parts of the structure.
   C. Have the member with the detector start his investigation at the heating source.
   D. Allow the occupants to return after venting the area and testing to find less than 35 ppm of CO.

69. All but one of the following are correct statements regarding fires in fireproof high-rise apartment buildings. Which does not belong?

   A. When confronted with high-wind conditions, you must evaluate the effect that venting the windows will have.
   B. If venting from inside the fire apartment under these conditions, make a small experimental opening and see what effect it has on the fire.
   C. If venting from the floor above, be sure that the door to the stairwell is closed.
   D. If the wind blows in when the windows are open, it's best not to vent until the hoseline has thoroughly cooled the area.

70. As the incident commander, you arrive simultaneously with the first engine at a working fire in an old downtown three-story brick and wood-joist building. The fire seems to be located in an office supply store on the first floor, with no obvious extension to the occupied apartments above. As the engine stretches its line to the front door, you note that the show windows are pitch black, the glass is cracked, and heat is radiating from them. There is no open flame visible, but heavy smoke puffs from around the door, then seems to be drawn back inside. At this point, which of the following would be most appropriate?

A. Order the engine company to take a position opposite the fire building for possible use of its master stream.

B. Order the show windows vented immediately after the engine gets water in its line.

C. Order the engine company to use an indirect attack with a wide fog pattern after the show windows are vented.

D. Order the ladder company members to withhold venting the upper floors until the ground-floor fire has been extinguished.

71. An engine company officer arriving to find heavy fire in control of the front 25 percent of a 200-foot-long by 50-foot-wide lumber storage building should know that, to knock down this much fire, the crew will have to attack with most nearly _____.

A. 250 gpm.
B. 1,250 gpm.
C. 2,500 gpm.
D. 25,000 gpm.

72. As the officer in command of the roof sector, you are informed that the roof you're working on appears to be a rain roof. Given this bit of information, you would be most correct in which of the following?

A. The stability of the roof is not a major concern if, when you make an inspection opening, there is no fire directly under your position.

B. The biggest problem will be in determining the extent of the fire in the cockloft.

C. Roof ventilation will be very rapid, given the nature of the roof.

D. Once the rain roof is open, you should push down the ceilings below to expose the cockloft.

73. Which choice is not true regarding lightweight parallel chord trusses?

A. They create large open spaces for fire to travel in.

B. They can be recognized by their classic hump shape.

C. Operating a power saw on the deck that they support may cause total failure of the truss.

D. The failure of a single web member or any of the connectors can cause total truss failure.

74. Which of the following would not be an acceptable tactic for a fire in a row of garden apartments if the building is comprised of lightweight parallel chord trusses as floor and roof supports?

A. Operate from the perimeter, using the reach of the stream on the contents.

B. Provide total ventilation directly over the fire.

C. Open the side wall from a platform and direct a stream into the truss loft.

D. Advance slowly after the fire has been knocked down, and use adequate lights.

75. A firefighter finds himself suddenly cut off by extending fire. Which of the following actions would not be in his best interest?

   A. The firefighter stays calm and stays put to conserve air.
   B. The firefighter calls for help by all means possible.
   C. The firefighter closes doors between himself and the fire.
   D. The firefighter activates his PASS device.

76. A chief officer arrives at the scene of a large building that has collapsed pancake-style. Which of the following orders is inappropriate, given the situation?

   A. The chief orders the first- and second-arriving engines to take hydrants in the block and supply handlines or master streams.
   B. The chief orders aerial ladders to take positions away from the front of the collapse site.
   C. The chief orders an elevating platform to be positioned in front of the collapse site, outside of the secondary collapse zone.
   D. The chief orders that space be reserved near the collapse scene for several heavy-rescue units, which he calls for special equipment.

77. Which of the following hydraulic principles is incorrectly stated?

   A. Regardless of the diameter of the hose, the more water flowing through it, the higher the friction loss will be.
   B. Regardless of the length of the hose, at a given flow, the friction loss remains the same.
   C. Regardless of the length of the hose, if there is no flow, the pressure remains the same throughout the line.
   D. Regardless of the flow, by increasing the diameter of the hose, the same amount of water can be moved farther with the same amount of energy.

78. The captain of a ladder company, while conducting training in the quarters he shares with an engine company, makes the following statements regarding engine company operations. In which choice does the captain misstate a fact?

   A. When positive pumping through large-diameter hose, the pumper stops at the fire and drops off a variety of nozzles, hose, manifold, and appliances, then proceeds to a water source.
   B. By placing the manifold near the fire, a variety of options is available.
   C. A manifold might be used to supply handlines, master streams, or other pumpers.
   D. The manifold should supply only one class of device, but if that means a handline, then there is no problem.

79. When using the combination attack on a free-burning fire involving several rooms of a home, which would not be a proper action for the nozzleman to take?

   A. To use a very narrow fog pattern directed at the ceiling.
   B. To move the stream in a side-to-side sweeping motion along the ceiling.
   C. To continue this sweeping motion along the ceiling as the line is advanced from room to room.
   D. After the fire has darkened down, to shut down the nozzle to give the smoke, heat, and steam a chance to lift.

80. An engine company officer arrives to find one three-story frame building fully involved with fire, and a similar building located thirty feet upwind smoking from the radiated heat. Which would be the most correct action for the officer to take?

    A. Order the driver to position the apparatus to use the apparatus deck pipe on the fire building.
    B. Use the deck pipe to place a water curtain between the fire building and the exposure.
    C. Have a handline positioned to coat the fire building with water to absorb the radiant heat.
    D. Use another handline with a solid-tip nozzle to assist the deck pipe in cooling the exposure windows, since the handline could get very close to the building.

81. A ladder company officer is directed to supervise the operations on the roof of a well-involved commercial building. On reaching the roof, he would be correct in ordering an immediate evacuation if he discovered that it was of which type?

    A. A 2 × 4 inverted roof with a ¾-inch deck.
    B. A standard sawn-joist roof with 1 × 6 roof boards.
    C. A metal deck roof on protected steel I-beams.
    D. A plywood roof deck supported by wooden I-beams.

82. All but one of the following are correct actions to take when cutting a trench on the roof of an H-shaped apartment house. Which does not belong?

    A. The trench is cut close enough to the fire to limit extension, yet far enough away so that it can be completed before the flames advance past it.
    B. The trench should be cut where the building profile is at its widest.
    C. The trench is located in the path of anticipated fire travel.
    D. The trench must be reinforced with hoselines.

83. All but one of the following descriptions of a team search are correct. Which does not belong?

    A. Team search needs a leader and either two or four searchers, plus a control person at the entrance and a two-person rescue team.
    B. The control person pays out the desired amount of rope as the team advances.
    C. The searchers are attached to the main line by 25- to 50-foot lengths of rope.
    D. The searchers fan out from the main line and move forward until they meet.

84. What are Norman's three rules of survival on the fireground?

    1. Never put yourself in a position in which you are depending on someone else to come get you out.
    2. Never go above the fire.
    3. Always know where your escape route is.
    4. Always have a charged hoseline available before entering a burning building.
    5. Always know where your second escape route is.

    A. 1, 2, and 3.
    B. 1, 2, and 4.
    C. 1, 3, and 5.
    D. 1, 3, and 4.

85. When establishing the protocols for a rapid intervention team, four items are crucial to success. Which of the choices below is incorrectly stated?

    A. People—You must have sufficient numbers at the scene to do the job. At minimum, you must have two for most trapped-firefighter scenarios.
    B. Policy—A plan for the response, use, and release of the RIT to ensure that it accomplishes its goals.
    C. Tools—Each RIT member should have certain basic equipment, and the team should have designated tools as well.
    D. Techniques—RIT members should have practiced the methods to be used to remove an unconscious firefighter.

86. A chief officer, arriving at a working fire in a newer strip mall, finds fire venting from the end store in a row of fourteen stores. The roof team reports that the roof is sagging over the fire store. What would be the most correct order for the chief to give?

    A. Advance a handline into, and vent the roof over, the involved store.
    B. Advance a handline into an exposed store ahead of the fire, and begin sweeping the cockloft with the stream. Have the roof team drop back to this store.
    C. Evacuate the roof as well as the interior of the building, and conduct only exterior operations.
    D. Evacuate the roof only, and advance the handline into the fire store, sweeping the ceiling.

87. Which of the following statements is untrue about stack effect?

    A. Stack effect is most pronounced in tall buildings.
    B. Stack effect during a top-floor fire on a cold day can be an extreme problem.
    C. Stack effect during a lower-floor fire on a cold day can be an extreme problem.
    D. Stack effect during a top-floor fire on a hot day can be an extreme problem.

88. An engine company arrives at a private home to find a strong odor of gas in front of the building and a hissing sound coming from the peck vent. The members should take all but which of the following actions?

    A. Notify the utility company of a possible regulator failure.
    B. Shut off the gas, at the curb cock, if possible.
    C. Pull the electric meter to remove sources of ignition.
    D. Vent the building and search as needed.

89. Early on a still summer evening, your company arrives first at a crowded new residential development. You find a fire in a barbecue involving a 20-pound propane tank in the rear of one of the homes. Which of the following orders is ill-advised?

    A. Secure a water supply and stretch a 1¾-inch line to cool the cylinder.
    B. Have the second engine stretch another line to the interior of the home to check for extension there.
    C. Evacuate the homes on each side and to the rear.
    D. Once the lines are in place, order the nozzleman to extinguish the fire and then shut off the cylinder valve.

90. A chief officer in command of roof operations at a large fire in the cockloft of an H-type building is supervising a trench-cut operation. In which of the following orders would he be correct?

  A. Make the trench two feet wide from fire-wall to firewall.
  B. Make sufficient inspection holes on both the fire side and the safe side of the trench.
  C. The trench must be subdivided into two-foot segments.
  D. Don't begin pulling the trench until fire shows at the inspection holes on the safe side of the trench.

91. A chief officer arrives at a fire in a high-rise residential building to find the wind blowing the fire back into the building. On opening the door from the stairway into the public hall, the firefighters are forced back down the stairs by flame. Under these circumstances, what should be the last thing the chief orders his company to do?

  A. Simultaneously advance with two hand-lines from the same stairway.
  B. Breach a hole from the stairway into the adjacent apartment, and advance within it to a point where they can put water on the fire.
  C. Use PPV fans to push the fire back into the apartment and out of the windows.
  D. Use an outside stream to darken down the fire.

92. At an incident in an older supermarket, the high heat and heavy smoke from the cockloft in the rear of the store prevent handlines from advancing to the seat of the fire. The chief orders the lines withdrawn to the sidewalk in anticipation of full involvement. Under the circumstances, the chief's actions were:

  A. Appropriate, because the inability to advance makes full involvement inevitable.
  B. Inappropriate, because the sidewalk is within the collapse zone.
  C. Appropriate, because the handlines will be operating in better conditions outside.
  D. Inappropriate, because the lines should have been relieved with fresh personnel and advanced to the seat of the fire.

93. Your unit is part of the assignment dispatched for an odor of smoke in a department store just after closing time. On arrival, you are met by the building manager, who tells you that the strong odor exists on all three floors of the building. Investigating, you smell the odor but detect no visible smoke. What is the most correct action for you to take as the ranking officer at the scene?

  A. Return all units and advise the owners to call back if they see smoke.
  B. Survey incandescent light fixtures for overheated ballast.
  C. Move to where the odor was last detected and examine appliances and devices in that vicinity.
  D. Send a team to the machinery rooms and examine the HVAC and elevator motors for overheating.

94. Which of the following statements concerning the behavior of steel when exposed to fire is the least correct?

    A. Steel is a brittle material that can fail suddenly.
    B. When heated to about 1,000°F, a 100-foot steel beam will expand almost ten inches.
    C. When steel reaches 1,500°F, it begins to fail.
    D. When hot steel is hit with water, it freezes in shape and returns to its original length.

95. A ladder company officer arrives at the scene of an early-morning fire in a wood-frame apartment building, one that has a single apartment on each of its three floors. Fire is venting out of two windows on the front of the second floor. Due to another nearby multiple-alarm fire, there will be a long delay in the arrival of the second ladder company. Which of the following choices is inappropriate under these circumstances?

    A. Transmitting calls for additional assistance.
    B. While searching the fire apartment, beginning the search at the apartment door, then penetrating as close to the fire as is safe.
    C. After completing the primary search of the fire floor, ascending to the floor above to search it.
    D. Beginning the search on the top floor at the apartment door and moving toward the front of the apartment, where there is a fire escape.

96. A chief officer is faced with a serious fire on the top floor of a four-story apartment house measuring 75 × 100 feet. The fire appears to have extended into the cockloft, but there are no adjacent exposures. The department's sole aerial device, a 75-foot snorkel, is heavily engaged in rescuing trapped occupants. Which of the following calls for assistance is the least correct?

    A. The chief calls for an additional snorkel to rescue other trapped victims in the rear.
    B. The chief calls for an additional aerial ladder to provide access to the roof.
    C. The chief calls for a ladder tower for possible use of its heavy stream.
    D. The chief calls yet another snorkel to the rear to act as an escape route for the many firefighters involved in roof operations.

97. Which of the following is an incorrect statement regarding roof cutting operations?

    A. The first consideration when deciding where to cut is the direction of the wind.
    B. You should plan an escape route in case conditions deteriorate.
    C. You should never cut so that a member has to step on a compromised portion of the roof.
    D. You should cut close to the inside of the joists when making the traditional 8- × 8-foot hole.

98. Which of the following is incorrect regarding hoseline positioning?

    A. For fires above grade, you should stretch dry to a safe area, usually via the interior stairway.
    B. For fires above grade, it may be beneficial to haul the line up and onto the fire floor by using a rope.
    C. For fires below grade, you should always charge and bleed the line before starting down the stairs.
    D. For fires in one-story buildings, you should try to attack from the unburned side if this will cut off the fire.

99. Which of the following is most correct regarding positioning a ladder apparatus for the most effective use?

   A. You should slow down before reaching the building to observe the conditions. If a rescue is needed, spot at the best location.

   B. If the device is mounted amidships, you should spot the apparatus at about a 45-degree angle away from the building.

   C. For rear-mounted devices, the best scrub area is obtained by nosing the apparatus in toward the building.

   D. If you can see no obvious need for the device while approaching the fire building, you should stop the rig about fifteen feet short of the far end of the building.

100. All but one of the following are safety precautions that you should take at the scene of a structural collapse. Which choice does not belong?

   A. Control the spread of fire. If no fire is present, prepare for it anyway.

   B. Eliminate vibrations from nearby highways, rail lines, nonessential apparatus, etc.

   C. Use all available personnel for maximum impact.

   D. Monitor the atmosphere for oxygen, flammable gases, and toxins.

PART IV

ANSWERS

Answers

1

General Principles of Firefighting

ANSWERS

1. Human life shall take precedence over all other concerns (pages 7-8).

2. B (page 8).

3. C (page 10).

4. D (pages 9-10). Note: Remove those in greatest danger first, if they are savable.

5. D (pages 10-11).

6. B (page 11).

7. C (page 12).

8. C (page 12).

2

Size-Up

ANSWERS

1. A size-up is a continuing evaluation of all the problems and conditions that affect the outcome of the fire (page 15).

2. Size-up should be performed by all members to varying degrees (page 15).

3. Size-up should be begun on receipt of the alarm and continue until the incident has been brought under control (page 15).

4. D (page 15-16). Note: All of the remaining items may or may not be true and must be verified on arrival.

5. Verify the answer in your own town.

6. C (page 16).

7.   1. C — Construction.

    2. O — Occupancy.

    3. A — Apparatus and manpower.

    4. L — Life hazard.

5. W — Water supply.

6. A — Auxiliary appliances.

7. S — Street conditions.

8. W — Weather.

9. E — Exposures.

10. A — Area.

11. L — Location and extent of fire.

12. T — Time.

13. H — Height (page 17).

8. Combine Area and Height and add Hazardous materials (page 17).

9. D (page 17).

10. By installing a complete wet-pipe automatic sprinkler system (page 18).

11. (1) The potential life hazard, (2) the presence of large open-floor spaces or small rooms, (3) the presence of hazardous materials, (4) the degree of fire loading, and (5) the possible presence of truss construction (page 18).

12. C (page 20).

13. A (pages 20-21).

14. (1) Class I, fireproof (most collapse resistant), (2) Class III, heavy timber, (3) Class IV, ordinary, (4) Class V, wood frame, (5) Class II, noncombustible (least collapse resistant) (pages 21-23).

15. (1) Building built on a grade, (2) wraparound building, and (3) interconnected buildings (pages 24-25).

16. D (pages 25-26). This fire is not below grade, not out of the reach of ladders, and not in a windowless building.

17. (1) Location, (2) color, and (3) movement (pages 26-27).

18. (1) 10 gpm/100 sq. ft., (2) 20 gpm/100 sq. ft., (3) 30-50 gpm/100 sq. ft. (page 30).

19. (1) Sprinkler system, (2) standpipe, (3) foam system, (4) CO_2 flooding system, (5) range hood dry chemical system, and (6) halon flooding system (pages 31-32).

20. (1) Fatigue firefighters, (2) slower operations; mechanical failures, and (3) drive attack crews off the floor; past defensive measures (page 32).

21. B (page 33).

3

Engine Company Operations

ANSWERS

1. D (page 35).

2. (1) Direct attack, (2) indirect attack, and (3) combination attack (page 36).

3. Direct attack (page 37).

4. C (pages 37-38).

5. D (page 38).

6. B (page 38).

7. C (page 39).

8. D (page 42).

9. D (page 45).

10. D (page 45).

11. Offensive attack (page 46).

12. (1) Call for additional ventilation opposite the line, (2) call for an increase in pump pressure on the first line, and (3) call for a backup line (page 48).

13. (1) Change the direction of the attack, (2) use master streams to darken heavy fire, (3) breach a wall to darken the fire, and (4) use distributors, cellar pipes, high-expansion foam, etc. (pages 48-49).

14. D (page 50)

15. B (page 52).

4

Hoseline Selection, Stretching, and Placement

ANSWERS

1. (1) Occupancy, (2) construction, (3) height and area, and (4) the location and extent of fire (page 54).

2. A (pages 54-55).

3. C (pages 54-55).

4. D (pages 56-57).

5. D (page 57).

6. B (page 59).

7. B (pages 59-60).

8. B. 600' - 200' + 200' (length & width), plus two lengths for the two flights of stairs, plus 100' from hydrant to entrance (pages 59-61).

9. A. At least one length for the apartment, one length up the open stairwell (pages 62-64).

10. D. The first line must go to the interior stairs on the fire floor, just as a line coming up the staircase would (page 65).

11. D (page 71).

12. A (page 71).

13. B (pages 73-75).

14. B (page 73).

15. C (page 80).

16. A (page 82). Note: The area over the fire must also be vented.

17. C (page 84).

5

Water Supply

ANSWERS

1. C (page 88).

2. C (page 88).

3. B (page 91).

4. D (pages 90-109).

5. D (pages 96-99).

6. B (page 93).

7. C (page 95).

8. C (page 99).

9. B (pages 102-103).

10. C (pages 110-111).

11. B (page 113).

12. C (page 90).

13. B (pages 91-93).

14. C (page 92).

15. B (page 93).

16. D (pages 93-94).

6

Standpipe and Sprinkler Operations

ANSWERS

1. C (page 115).

2. A (pages 115-116).

3. A (page 116).

4. (1) A combination of hoselines and sprinklers may be holding the fire in check, or (2) a second fire may be burning in a remote area (page 116).

5. C (page 116).

6. B (page 117).

7. It will cause the sprinkler operating pressure and discharge to drop (page 117).

8. Stretch one of the first hoselines, supply the siamese, and back it up with a second line (pages 117-118).

9. B (page 117).

10. C (page 118).

11. C (page 118).

12. C (page 118).

13. B (page 118).

14. B (page 118).

15. Sprinklers that begin to operate after fire department operations have begun (page 119).

16. In buildings that are only partially sprinklered, where the fire begins in an unsprinklered area (page 119).

17. They negate venting due to their spray pattern (page 119).

18. (1) Preplan the location of shutoffs and siameses, (2) supply siameses early with proper volume and pressure, (3) stretch handlines to the seat of the fire, (4) vent the area as required using all necessary means, and (5) after the fire is definitely under control, shut down the sprinklers and restore protection (page 147).

19. A (page 121).

20. A (page 122).

21. In areas subject to freezing (page 123).

22. A (pages 123-124).

23. A (page 123).

24. C (pages 126-127). Note: Deluge systems use open heads throughout the area. Preaction systems have closed heads—only those over the fire discharge.

25. C (page 126). Note: (A) Deluge system. (B) Dry-pipe system. (D) Wet-pipe system.

26. D (page 128). Note: A, B, and C are all useful outward signs if D hasn't been done.

27. D (page 134).

28. A (pages 129 and 137).

29. Run two supply lines of the largest hose available to each siamese, and run supply lines into the first-floor hose outlets (pages 138-139 and 149).

30. C (pages 139-140).

31. B (pages 140-141).

32. Reflex time is the time required to respond to the building, proceed to the fire area, and connect or stretch hoselines prior to beginning the attack—i.e., the time from the alarm to the time that water is first applied to the fire (page 142).

33. Sufficient hose, at least three lengths of the proper diameter, nozzle (with a solid tip available), hose thread adapters or reducers as required, spanner, a 10-inch to 14-inch pipe wrench, door chocks, forcible entry tools) (flathead ax, halligan, K tool, handlight, portable radio, door latch straps (pages 145-147).

34. B (page 148).

35. (1) 65 psi in pre-1993 buildings, (2) 100 psi in post-1993 buildings with 1½-inch outlets, and (3) 175 psi in post-1993 buildings with 2½-inch outlets (pages 134-139).

7

Ladder Company Operations

ANSWERS

1. L — Laddering.

   O — Overhaul.

   V — Ventilation.

   E — Entry (forcible).

   R — Rescue (search and).

   S — Salvage.

   U — Utilities (control of) (page 151).

2. A (page 152).

3. (1) The number of personnel initially assigned to each major sector, (2) the tools to be provided, and (3) the general scope of duties (page 152).

4. C (pages 152 and 155).

5. It must be realistic (page 152).

6. For responsibility and tracking purposes, with another person working in the same general area (page 153).

7. The experience level of the member; the ability to act independently and protect himself; the availability of others in nearby positions who will be able to assist him as needed; the presence of a portable radio for this member (page 153).

8. One- and two-family dwellings (more than 70 percent of fire deaths) (page 155).

9. D (pages 156-157).

10. Vent, enter, search (page 156).

11. D (page 157). Note: The bedroom has the highest life hazard despite the time of day.

12. By assigning the tools and responsibilities to the riding position, so that whoever rides in that position will know his duties (page 158).

13. Reach, weight, stored length, material of construction, and the number of people needed to place it in operation (page 159).

14. The ability to adjust the ladder to the proper working height (page 160).

15. C (pages 162-163).

16. By completely clearing the laddered window of glass and sash (page 163).

17. D (page 163).

18. C (page 164).

19. A clear line-of-sight path from the turntable to the objective (page 168).

20. At least five rungs (page 168).

21. D (page 170). Note: A telescoping platform must be raised before retracting it.

22. B (page 173).

23. C (page 174).

24. About fifteen feet past the near edge of the building so that the apparatus may be driven forward if the need arises later. Position with the anticipated travel of fire in mind (page 175).

25. Any action taken to expose hidden fire and ensure its extinguishment (page 175).

26. Opening up any void to get ahead of a spreading fire (page 176).

27. Sight, hearing, touch, smell, and common sense (page 178-179).

28. Use the fifteen-second, two-minute rule (page 179).

29. In the doorway of the escape route (page 181).

30. At a seam between two sheets, or at an opening around a light fixture or pipe (page 181).

31. D (page 182). Note: (A) Vacant, no water damage. (B) Do not remove a lintel supporting a brick wall, since collapse may result. (C) Seriously damaged trusses pose severe dangers to firefighters.

32. Salvage is an effort made to reduce the damage to the remainder of the structure and its contents (page 182).

33. Examine both sides of the partition for the fastest, least damaging location to open it (page 183).

34. By not throwing water at smoke, and by draining the hoseline outside or into a sink or tub (page 183).

35. (1) Send two members at all times, and equip them with SCBA and a radio, (2) send someone with expertise in the area—i.e., electrical background, plumbing background, etc, (3) call for utility personnel to assist, (4) if in doubt of an action, seek further advice from experts (pages 184-185).

8
Forcible Entry

ANSWERS

1. A (pages 190-192).
2. A (pages 187-188).
3. D (pages 188-190).
4. B (page 192).
5. Whether the door opens in or out (page 192).
6. The outside radius toward the door (page 193).
7. D (pages 193-195). Note: All of the others destroy the integrity of the door.
8. The top hinge first, so that venting smoke or fire will go above the members as they crouch to do the lower hinge (page 195).
9. The adz end, with the fork facing back across the door (page 196).
10 A (page 197).
11. Pick up the cylinder, examine it, and determine what key tool to use (page 199).
12. An inward-opening door (usually residential) (page 200).
13. C (page 202).
14. B (page 205). Note: The cylinder of a fox lock is recessed in the door.
15. Toward the lower set of bolts at the door's edge (page 206).
16. If the visibility is poor (and if heavy fire is present or there is a need for speed) (page 206).
17. Determine whether the multilock is in fact engaged (page 210).
18. By looking for a rod between the door and the jamb (on outward-opening doors), or using a thin object to slip under and along the edge of the door (page 210).
19. D (page 211).
20. A knowledge of padlocks and how to remove them (page 211).
21. A (page 212).
22. A (page 213).
23. Cut the U channel to which it is secured (page 213).
24. D (page 214).
25. A (page 215). Note: Aluminum oxide blades are required for either the gate or padlock.
26. A (page 215).
27. D (pages 215-217).
28. B (page 218). Note: (A) Since you must be inside to operate series locks, the through-the-lock method wouldn't be possible. (C) The key from the previous lock in the sequence is required, not just from any lock.

9
Ventilation

ANSWERS

1. Ventilation is the process of removing the products of combustion from a structure by controlled means and replacing them with fresh air (page 220).
2. B (page 220).
3. The timing of the ventilation (page 220).
4. B (page 220).
5. A (page 220-221).
6. Opposite the hoseline (page 221).
7. An immediate entry for rescue or placement of a hoseline between the victim and the fire (page 221).
8. D (pages 221-222). Note: Especially wind.
9. D (page 222).
10. Top-floor, attic, and cockloft fires (page 222).
11. B (page 223). Note: Mechanical ventilation may be useful, but given the description of the fire, it wouldn't normally be a prerequisite.
12. B (page 223).

13. With mechanical ventilation, it's possible that the fresh air being drawn into the fire area may fan a smoldering fire into more serious proportions (page 225).

14. D (page 224).

15. B (page 225).

16. D (page 227).

17. Vent directly over any vertical arteries by which the fire is traveling (page 228).

18. Break a single, small pane first, then pause a few seconds to signal the members below to protect themselves (page 228).

19. B (page 231).

20. Hard roof coverings are extremely slippery when they're wet. Loose tiles or slate pose a falling-object hazard to people below, and they compound the fall hazard for firefighters (page 233).

21. D (page 234).

22.

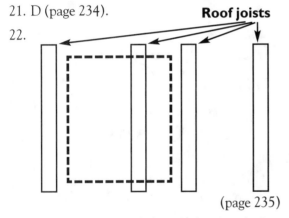

Roof joists

(page 235)

23. (1) Ensure the stability of the area before entering, (2) always have two means of escape, (3) plan the cut and inform other members of its layout, (4) keep the wind to your back, (5) cut adjacent to the joists, (6) never step on the cut, and (7) don't cut the roof supports (page 236).

24. Standard flat roof (page 237).

25. Inverted roof (page 237).

26. Four members, two power saws, two hooks, halligan tool, axe, and a portable radio (pages 239).

27. D (pages 228-230 and 239). Note: If staffing permits, choice 3 may be done simultaneously with choices 1 or 4.

28. A kerf cut is a small narrow opening that is only the width of the blade or the chain of a power saw. It is used to verify the presence and location of fire below (page 239).

29. B (pages 239-240).

30. D (page 241).

31. C (page 242). Note: Cut G is not required, but it should be made anyway to provide flexibility in case the wind shifts.

32. C (page 239).

33. A (pages 238 and 246). Note: There are no trusses in a standard roof.

34. A (page 244).

35. No, evacuation is not necessary. An inverted roof is relatively stable under fire conditions. It is entirely different from 2 × 4 truss construction, in that roof failure occurs locally in sections after a warning sag has developed. Even if it does fail, it will only drop down to rest on the main roof joists, which will usually be intact (page 244-246).

36. A rain roof is a new second roof built over the existing leaky roof. Often it is built on a raised frame similar to that of an inverted roof (pages 245-246).

37. C (page 246).

38. C (page 246).

39. C (pages 247-248).

40. Top chord, bottom chord, and web members (page 248).

41. C (page 249).

42. The classic humpback design in occupancies requiring large, open floor spaces without columns (page 249).

43. Their lack of mass to resist fire (page 251).

44. If power saws are used to cut the deck, they may sever the top chord, precipitating collapse (page 251).

45. They create large, open spaces beneath every deck. If there is a ceiling below, there is a truss loft on each level (pages 251-252).

46. Plywood I-beams show no noticeable sag before they suddenly fail (page 253).

47. D (page 249).

48. To create a fire break in the roof's surface so as to prevent fire from passing from one wing of a building to another (page 257).

49. D (pages 257-259).

50. D (page 258).

51. D (pages 538-540).

52. Their fellow firefighters, particularly those with power saws (page 263).

53. To stop the saw blade from spinning when it isn't in a cut (pages 263-264).

54. Steel plating added to the roof, walls, or ceiling (page 264).

10
Search and Rescue

ANSWERS

1. Primary and secondary (page 269).

2. The degree of fire control. This dictates how quickly you must move and how thorough you can be (page 269).

3. Primary search (page 269).

4. Secondary search (page 269).

5. B (page 273).

6. A (page 273).

7. D (pages 273-274).

8. A (page 274).

9. Cribs usually have high legs and barred sides (276-277).

10. A bunk bed has an unusually low bottom mattress, and it is confined on at least one end or side (page 277).

11. C (page 277-278). Note: A two-person team is required for each apartment or moderate-size store or office under light smoke conditions. More are required if the smoke is thick or if the searchers encounter heavy stock or a cluttered environment.

12. D (page 278).

13. No. The rope is intended to provide a reference point, fending off disorientation. It may be held by one member while others work nearby without being tied off, as long as smoke and visibility permit (page 279).

14. B (pages 279-281).

15. The rope must be secured at each change of direction (page 282).

16. It can help you see through smoke and mist to locate a victim (page 284).

17. (1) Feedback-assisted rescue, (2) pager-assisted rescue, and (3) PASS devices (pages 282-283).

18. D (page 283).

11
Firefighter Survival

ANSWERS

1. C (page 289).

2. D (page 291).

3. D (pages 291-292).

4. The reduced-profile and the quick-release procedures (pages 292).

5. D (page 292).

6. D (page 293).

7. B (pages 293-294).

8. C (page 294). Note: Do not walk! Lie with your face on the floor and your lights off momentarily.

9. D (pages 292-296).

10. C (pages 296-297). Note: A belay is only used in training.

11. D (page 297). Note: Post the second copy on the dash, and give the third copy to the command post.

12. B (page 298). Note: (A) In the scenario, the member has already been reported as missing. (C) Having a hoseline is good, but it might not be needed. Information is always required. (D) The member may be buried, his PASS device may be unarmed, or he might be okay and just unaccounted for.

13. A (page 299).

14. C (page 300).

15. D (pages 299-300).

16. A (pages 300-301).

17. B (page 301). Note: At least until the fire is under control.

18. A (page 302).

19. D (page 302). Note: (A) One 200-foot guide rope for each two-member team. (B) One set of forcible entry tools for each two-member team. (C) One spare mask for the trapped firefighter.

20. A (page 302). Note: The RIT personnel bring this with them. The rest they get at the scene.

21. D (page 303).

22. C (page 303).

23. A (page 304). Note: It should be an ALS, not a BLS.

24. D (page 304). Note: There is no guarantee that he is conscious or able to assist.

25. A (page 304).

26. B (pages 304-305).

27. C (page 305).

28. B (306).

29. B (page 306).

30. A (page 306).

31. D (page 307).

32. A (page 308). Note: (B) Use a backboard on stairways. (C) Secure the rope to the victim and the backboard. (D) One member stabilizes the head and neck and maintains an open airway. Also, the team uses a backboard, not a ladder.

33. C (pages 308-309).

34. C (page 309).

12
Private Dwellings

ANSWERS

1. C (page 311).
2. C (page 311).
3. A (page 311).
4. B (page 311).
5. C (page 312).
6. C (page 314).
7. C (page 314).
8. (1) An in-line stretch permits the fastest water application possible, using preconnected lines off the booster tank; (2) it ensures a continuous water supply; and (3) it makes a master stream available in front of the building (page 312-313).
9. C (page 314)
10. A (page 314). Note: (B) Also requires a line on the first floor. (C) Poke a small examination opening in each bay; don't bother with the whole height initially. Don't bother with window and door bays, since they act as firestops. (D) Requires roof venting.
11. B (page 316).
12. C (page 313).
13. B (page 314).

13
Multiple Dwellings

ANSWERS

1. B (page 317).
2. D (page 317).
3. D (page 317).

4. C (page 318).

5. Single-room occupancy. Persons each rent a single room for sleeping, sharing kitchen and toilet facilities with others in the building (page 318).

6. A (pages 318-319).

7. A (page 320).

8. C (pages 320-321). Note: There are often no visible indications of where this space is located.

9. B (pages 320-321). Note: Send someone who has seen this shaft and knows where to look, if possible. Be sure to send someone to the top floor as well. Don't forget the base of the shaft. Hose streams operating in the shaft delay extension downward.

10. B (pages 320-321). Note: (A) Only remove those in great danger. Protect the rest in place. (C) Don't waste time and personnel when putting out the fire will solve the same problem.

11. A (page 322).

12. D (page 323).

13. B (pages 323-324).

14. Have a member proceed to the second floor and locate the proper apartment line (page 324).

15. C (page 324).

16. A (page 324).

17. Return stairway (page 325).

18. C (page 325).

19. B (pages 326-327).

20. B (page 328).

21. It provides a stable work platform, it allows for close observation of the master stream, and it permits extremely accurate operation of the stream in all directions through an opening (page 328).

22. B (pages 328-329).

23. D (page 329).

24. The members of the roof team should descend to begin searching above the fire, either by way of the fire escape or by returning to the street and then making their way up through the interior (page 330).

25. First, vent the top-floor windows from the roof. Second, begin roof cutting. If descent via the fire escape is possible, venting and search from this area is preferred. Then, cut the roof (pages 330-331).

26. Drop ceilings create cocklofts on each floor that are interconnected with the vertical pipe chases, a maze for fire to travel in (pages 331-332).

27. (1) Deterioration due to weather and previous fires, (2) lightweight construction, and (3) cornice collapse (page 332).

28. B (pages 332-333). Note: Collapse is not a consideration in occupied Class I buildings.

29. A (page 334).

30. B (page 334).

31. The occupants should stay in their own apartments (page 337).

32. Use the apparatus public address system (pages 337-338).

33. B (page 323).

14

Taxpayers

ANSWERS

1. Class IV (ordinary) construction housing several commercial occupancies under one roof—i.e., a row of stores (page 339).

2. A (pages 339-341).

3. B (page 340).

4. Trench cutting requires too much time to complete and would surrender an extremely large part of the building (page 340).

5. Order the roof forces to back away twenty-five to thirty feet from the danger area and resume operating. Pull the interior forces out of the involved area and operate in exposed stores

until the main body of fire has been darkened down and the steel has cooled (pages 341-342).

6. C (pages 340-343).

7. D (pages 343-346).

8. B (pages 344-345).

9. B (page 346).

10. D (page 347).

11. D (page 342 and photo on page 357).

12. D (page 348).

13. D (page 348).

14. Very inaccurate. A wide fog will produce a tremendous amount of steam, burning personnel, and it could push fire around behind the line (page 348).

15. D (page 348).

16. D (pages 348-349).

17. A (pages 349-351).

18. D (page 350).

19. B (pages 352-353). Note: (A) For a heavy fire, use a master stream in the fire store, and use 1¾- to 2-inch lines against the exposures. (C) More than one line is needed per store. (D) The stream should be applied from the level of the sidewalk.

20. D (pages 353-354).

21. D (pages 354-355).

22. The walls above show windows are carried on steel I-beams. These I-beams are often connected to other beams that run front to rear in the cockloft. Any twisting of the beam at the window, or expansion of the intersecting beams, can topple the parapet (page 356).

23. They can push through solid brick or cement-block walls, allowing the fire to spread (page 356).

24. The advantage of a single large line is that it has more reach and penetration with the same or less staffing required (page 357).

25. A (page 356).

26. B (page 348).

15

High-Rise Office Buildings

ANSWERS

1. A fire-resistive building that is windowless or out of the reach of ladders due to its height or setback from the street, and where all operations must be conducted from the interior (page 360).

2. D (pages 360-361).

3. D (page 363).

4. D (pages 360-361). Note: Evacuation isn't normally required initially, but it is essential for large fires.

5. C (page 361).

6. D (page 362).

7. B (page 362).

8. B (page 362).

9. C (page 363).

10. The incident commander must know the floor layout, and he must have plotted the location of the fire, all stairways, firefighters, and civilians, and he should have an idea of what will happen if the system is turned on (page 364). Note: The HVAC system should only be used for such purposes on the orders of the incident commander. The normal HVAC system is only useful on relatively small fires, where the heat won't melt fusible links on fire dampers and where the volume of smoke is relatively light. The use of the system will draw heat, smoke, and fire toward the return air collectors in the ceiling, and as such could be a threat to firefighters and civilians in the path of the gases.

11. Shut down the entire system (pages 363-364).

12. A (page 364).

13. Stack effect is the natural movement of air within tall buildings. It is caused by the warm air at each floor trying to rise to the top of the building as cooler air sinks (pages 364-368).

14. No, it happens normally all the time, but a fire may intensify its effects (page 364).

15. A (pages 365-366).

16. B (page 366).

17. Stratification and the reverse stack effect. Stratification occurs when there is cold smoke, such as occurs when sprinklers have operated. The smoke may not rise and may get stuck in a zone of equilibrium. Reverse stack effect occurs when the outside air temperature is much warmer than the inside temperature, and it may result in smoke moving down to floors below the fire. It is also more likely to occur with cooler smoke (pages 366-367).

18. C (pages 367-368).

19. If conditions are such that a high-rise fire involves lower floors when there are low temperatures outdoors, or involves high floors when the temperatures outside are high, then horizontal ventilation may cause air to flow inward, toward the stairway, rather than out the windows (page 368).

20. That the elevator will stop at the fire floor (page 369).

21. Reflex time is the time that passes from the receipt of the alarm until an effective hose stream is operating on the fire (page 369).

22. B (page 369).

23. B (pages 369-370).

24. This stream provides a high volume (325 gpm) and long reach at low nozzle pressure. The building pump is limited to 65 psi at the top-floor standpipe outlet, mandating a nozzle that operates at 50 psi or less (page 373).

25. In core construction, all of the nonrentable spaces, such as stairways, elevator shafts, and utility closets, are in the center portions of the floor layout, leaving the highly desirable perimeter for offices, cubicles, and open space. This enhances fire spread in the outer areas. Because the stairways are grouped together, core construction reduces the options from which an attack might be mounted, and it allows fire, heat, and smoke to wrap around personnel operating via the stairway (page 373).

26. B (pages 373-374).

27. Using a core drill to cut an access hole or using a Lorenzo ladder to apply a ladder pipe stream from the floor below the fire (pages 374-375).

28. Scissor stairs usually alternate from one side of the core to the other. Personnel operating on the floor below the fire will be on the opposite side of the building from the fire and will be familiarized with a floor plan that is different from the one that exists on the fire floor (pages 375-376).

29. B (page 376).

30. Either the stairway should be located within a fire-rated enclosure or all of the connected floors should be fully sprinklered, with a separate deluge sprinkler system protecting the access stairway. In all cases, the floors served by access stairs should be posted in the lobby for the benefit of firefighters (pages 376-377).

31. (1) Command post in the lobby, (2) operations post on the fire floor or the floor below, (3) staging area, approximately three floors below the fire, and (4) search and evacuation, above the fire (pages 377-379).

32. B (pages 377-378). Note: At the operations post, there should be plans for the fire floor and the floor above, not all the floors above. These should be left at the command post.

33. Arranging the necessary support to keep the attack lines moving forward is the major task of the operations officer (page 378).

34. At least two attack teams (engines) and two support teams (ladders) present; an EMS sector; sufficient storage space for these personnel; plus spare hose, tools, air bottle, etc. The staging area should be in the vicinity of the attack stairway, and it should have good communications with both the operations post and the command post on several channels. A member should be monitoring communications from both of these entities (pages 378-379).

35. The ability to get past the fire with a reasonable degree of safety and the ability to retreat if necessary. The ability to communicate easily with the operations post and the command

post, as well as with all the personnel who are operating as part of the search and evacuation effort above the fire (pages 379-380).

36. To direct and control the activity of all forces operating above the fire floor and the floor immediately above; to verify and record the results of searches on all the floors above the fire area; and to maintain contact with the operations post so that all personnel above the fire can be withdrawn to safety if the fire cannot be controlled (pages 379-380).

37. D (pages 378-379).

16
Buildings Under Construction, Renovation, Demolition

ANSWERS

1. A (page 381).

2. A (page 381).

3. B (pages 381-382).

4. A (pages 382-385 and 388-389).

5. B (pages 382-383).

6. B (pages 383-384).

7. B (pages 383-384).

8. D (pages 384-385).

9. The ceiling panels are among the last items to be installed, and they're the first taken out for renovations. The absence of these panels allows fire to attack the structure above (page 386).

10. Spray-on fireproofing is often scraped off to run utilities and for other construction purposes, thus threatening failure of the steel if a serious fire occurs nearby (page 386).

11. C (page 387).

12. C (page 387).

13. To protect the occupants of a building that is still under construction, install basic exposed pipe sprinkler systems on all the floors up to and below any occupied floors (pages 388-389).

17
Fire-Related Emergencies: Incinerators, Oil Burners, and Gas Leaks

ANSWERS

1. A (pages 391-392).

2. Methane and ethane (page 391).

3. D (pages 391-392).

4. The odorant can settle out downwind, making people believe they are in danger when in fact no gas is present (page 392).

5. The odorant may be filtered out by the soil, while the gas migrates into buildings and manholes. This can result in a flammable mixture with no gas odor (page 400).

6. A (page 392).

7. C (page 393).

8. A (page 393).

9. The excessive gas pressure can (1) cause either the pilot lights or the burner flames to increase greatly in size, igniting nearby combustibles, or (2) blow out the pilot or burner flames, allowing a buildup of unignited gas (page 394).

10. B (pages 395 and 398).

11. Manufactured gas contained much carbon monoxide (page 395).

12. Notify the involved utility (pages 395-396).

13. How strong is the odor, and when did it first appear? (Page 397)

14. A (page 397).

15. C (page 398).

16. B (page 398).

17. If the gas/air mixture is already well above the upper explosive limits (pages 399-400).

18. D (page 400).

19. The soil filters out the odorant as the gas follows utility pipes into the building (page 400).

20. D (page 401).

21. D (page 402).

22. B (page 404).

23. C (page 404).

24. Cool the upper vapor space of the container (page 405).

25. D (page 406).

26. To direct vapors away from sources of ignition and to dilute the vapor/air mixture with air that was entrained in the fog streams (page 407).

27. Atomization—i.e., breaking up the oil stream into small particles, thus providing a high surface-area-to-mass ratio (page 409).

28. Through the use of the emergency or remote control, and closing the fuel valve at the tank (page 411).

29. D (page 411).

30. Incinerators are built to contain fire, and their chutes are heavy and fire-resistive. Compactors are not meant to contain fire, and their chutes are lightly constructed and nonfireproof (pages 416-417).

18
Collapse

ANSWERS

1. C (pages 421-425).

2. Fire involving the plywood forms of a recently poured floor could precipitate collapse (page 422).

3. Fires in heavy-timber structures don't usually cause collapse until after the fire has become so fierce that it drives personnel out of the building. The exception is a building that has experienced multiple serious fires and could collapse from a later, less serious cause (page 422).

4. (1) Steel expands when it's heated, possibly pushing over walls and columns. (2) When heated to more than 1,500°F, steel sags and twists, dropping its load. (3) When cooled, steel contracts to its original length, possibly shrinking away from its supports and causing additional collapse (page 424).

5. In a framed structure, most of the weight is carried on a skeleton or framework of steel or concrete (pages 424-425).

6. The collapse of an unframed structure is often a more catastrophic event than that of a framed structure (page 425).

7. Cast iron is an extremely brittle material (page 426).

8. (1) Structural weakness, (2) fire damage to wood structural members, (3) heating of unprotected steel, (4) heating and cooling of cast iron, (5) explosions, (6) overloading of floors or roofs, (7) expansion and overloading of absorbent stock, (8) cutting or removing structural elements during overhaul, (9) vibration and impact load, (10) miscellaneous factors, such as wind, flooding, etc. (pages 425-428).

9. (1) Hazardous occupancies, such as plumbing supply companies and appliance dealers, (2) dangerous construction, (3) overloaded floors, (4) heavy fire present for more than twenty minutes (standard construction, not truss or lightweight), (5) lack of runoff water, (6) cracks or bulges in walls, (7) water or smoke seeping through brick walls, (8) roof pulling away from the wall, (9) roof sagging or feeling abnormally soft or spongy, (10) obvious movement, (11) noises, such as creaking, cracking, or rumbling, (12) plaster sliding off walls, windows cracking, doors swinging (pages 428-431).

10. The incident commander must contact each unit on the fireground, confirm their receipt

of the evacuation order, and then ensure that all members observe safe exterior collapse zones (page 431).

11. Have all apparatus on the fireground turn on their audible warning devices and leave them on until ordered to turn them off (pages 431-432).

12. At least as large as the length of the facing wall and greater than its height (pages 432-433).

13. Members operating aerial devices, particularly platforms (pages 432-434).

14. Flanking positions, off to the side of the dangerous wall, and positions above the height of the wall (page 435).

15. D (page 434).

16. (1) Reconnaissance or survey, (2) surface victim removal and accounting, (3) search of voids, (4) selected debris removal and tunneling, (5) general debris removal (page 439).

17. (1) What happened and where, (2) who and how many are missing, (3) where were they last seen, and can they still be alive, (4) what assistance is needed, (5) what is the danger of fire, explosion, or secondary collapse, (6) what is the construction of the building, (7) are there problem occupancies with chemicals involved, and (8) can the utilities be controlled? (Page 439)

18. The victim tracking coordinator should interview all persons being removed from the scene, noting the names, identifying features, injuries, where transported, and the means of transportation. By asking about who was nearby when the collapse occurred and how the victim escaped, searchers can be directed to the area via the fastest route. By keeping track of all victims leaving the area, the search can be halted as soon as all victims are accounted for, thus ending any threat to the lives of searchers (page 440).

19. C (page 441).

20. A (page 441).

21. B (page 441).

22. On the floor below the collapsed floor, against the standing wall (page 440).

23. Void searches are much faster, since no digging is required. They are also safer for the same reason. Victims in voids have a better chance of survival than do those buried in debris (page 442).

24. He should monitor the condition of the members, ensure relief at intervals sufficient to prevent fatigue, note their progress, and arrange for necessary support operations to ensure a smooth operation. He should not become involved in manual labor (page 443).

25. Staffing (particularly trained personnel), the availability of equipment, and the stability of the debris all determine the feasibility of multiple tunnels (page 443).

26. It is best to select a tool that produces no exhaust fumes, no noise, and no sparks, and that does the job as quickly as possible. Also consider vibration (page 443).

27. Shore it up, tie it off, or pull it over in a safe area (pages 443-444).

28. General debris removal is necessary to ensure that no victims have been overlooked—e.g., vagrants, passersby, etc. This operation should be performed at all collapses (page 444).

29. (1) Shut down all utilities, (2) monitor the atmosphere for flammable and toxic gases, as well as sufficient oxygen, (3) permit no smoking, (4) remove all nonessential personnel, (5) control the spread of fire, or prepare for fire if none is present yet, (6) use a surveyor's transit to monitor any movement in weakened walls and floors, (7) eliminate all vibration, (8) do not cut or remove major supports, (9) if you absolutely must cut a support, brace it, shore it, and prepare for secondary collapse, (10) rotate personnel frequently, every half-hour or less, (11) maintain communications between the rescue teams and between the rescuers and the victims, (12) seek expert assistance (page 447).

Final Examintion

ANSWERS

1. A (page 47).

2. C (page 40).

3. D (page 58).

4. B (pages 98-99).

5. C (page 326).

6. D (page 109).

7. C (page 121).

8. D (page 341).

9. C (pages 129-130). Note: (1) Not acceptable. The system has not reset. In a large building, you cannot be sure that fire or water damage is not occurring. (2) Not acceptable for a similar reason, but you need additional information—e.g., how long before the foreman arrives; consider the temperature and the damage being done. A closed valve on alarm piping would prevent the water motor gong from operating and prevent drain discharge, yet water would still be flowing inside. (5) Do not shut down the system until the cause for the alarm has been verified and the entire area searched to ensure that the fire is under control. These valves are then operated on command.

10. D (page 138).

11. A (page 163).

12. D (pages 139 and 140). Note: No. 4 depends on the floor of the fire.

13. B (page 26).

14. D (page 116).

15. B (pages 156-157).

16. D (page 160).

17. D (page 148).

18. B (page 171).

19. C (pages 177-178).

20. D (page 197).

21. C (page 208).

22. C (page 222).

23. B (pages 229 and 330-331). Note: (A) Not yet. Be sure there is no fire in the pipe chase or on the way up to the cockloft, and never descend via the interior stairway until after the fire is under control. (C) Not for a ground-floor fire. (D) Not unless fire is in the cockloft.

24. A (page 254).

25. D (pages 258-261). Note: The fire in choice C is in the cellar.

26. C (pages 269 and 275-277).

27. D (page 314).

28. D (page 323).

29. D (page 339).

30. C (page 346).

31. B (pages 349-351).

32. A (page 362).

33. C (pages 365-366).

34. C (pages 369-372). Note: (A) Fireman's service elevators are affected the same as all the others. (B) Avoid freight elevators. (D) After precautionary stops every five floors, go to two floors below the fire.

35. D (pages 386-387 and 421-422).

36. D (pages 401-402). Note: Do not touch any valves in the street, since this could worsen the overall problem.

37. C (page 235).

38. D (pages 67-68). Note: The fire is in the rear, the house is occupied, and there is no mention of exposure problems. The protection of the exits must receive priority to permit search and prevent interior fire spread.

39. A (page 318).

40. D (page 354).

41. A (pages 293-294).

42. C (pages 290-292).

43. C (page 298).

44. C (page 140).

45. C (page 222).

46. D (page 227).

47. D (page 249).

48. B (page 128).

49. C (pages 343-344).

50. B (page 344).

51. B (page 346).

52. B (pages 348-349).

53. D (pages 352-356).

54. A (page 394).

55. D (pages 397-401). Note: For utility company use only.

56. D (pages 250 and 435). Note: At least two to two-and-a-half times the height.

57. A (pages 81-82).

58. C (page 110).

59. C (pages 37-38).

60. C (page 12).

61. C (pages 436-438).

62. C (page 10).

63. C (page 431).

64. D (pages 25-26).

65. C (page 429).

66. B (page 103).

67. B (page 412).

68. A (pages 417-418). Note: (B) At least two members. (C) Start at the front door. (D) In a home, anything more than 9 ppm is a problem. Do not allow reoccupancy unless the source has been removed.

69. C (page 334). Note: All doors must be positioned the same as they will be on the fire floor. Otherwise, you won't get a true picture of whether the wind will blow in on the fire.

70. A (pages 42-43). Note: This third-stage fire with its backdraft potential occurs with the occupied apartments above. The potentially heavy fire demands the placement of a heavy-caliber stream for rapid knockdown. (B) If faced with potential backdraft, take out the show windows after the engine has bled its line, but have the personnel stand off to one side to avoid blast damage. Better yet, use the master stream from across the street to blow the windows in. (C) After the show windows

have been vented, there is no chance for an indirect attack. (D) This is an occupied building, so you must vent the upper floors as soon as possible while conducting searches.

71. B (page 30). 50' × 50' = 2,500 square feet involved × 50 gpm per 100 sq. ft. = 1,250 gpm.

72. B (pages 43 and 245-246). Note: (D) The original roof will prevent you from pushing down the ceiling.

73. B (page 251).

74. B (page 253).

75. A (page 293). Seek an escape route, do not sit still. That's for when fire is not chasing you.

76. A (pages 445-446).

77. B (pages 93-95). Friction loss increases as the length increases.

78. D (page 106). Note: Handlines of different diameters, lengths, and nozzle types have very different pressure requirements and could pose serious problems.

79. C (pages 39-40). After cooling the overhead for ten to fifteen seconds, you should lower the stream to cool the rest of the burning material in the room and sweep the floor of hot debris.

80. A (page 45). Note: (B) This answer puts the water on the exposure, not in a water curtain. (C) This answer coats the exposure, not the fire building. (D) Use straight streams from a distance only when hitting windows for exposure protection.

81. D (pages 240 and 252).

82. B (page 258). Note: Narrowest, not widest.

83. B (pages 278-281).

84. C (page 291). Note: (2) Firefighters have to go above the fire to search for victims and extension. (4) Not always possible or practical—e.g., a third-floor fire, stretch dry to a safe area.

85. A (pages 299-300). Note: The minimum for most trapped-member scenarios is four.

86. B (pages 342-343).

87. B (pages 364-368).

88. C (pages 395 and 398-399). Note: Do not pull the meters, since there could be gas in them that may ignite.

89. D (pages 405-407).

90. B (page 259). Note: (A) Three feet. (B) Four feet. (D) Fire side.

91. A (pages 334-336).

92. B (page 356).

93. D (page 415). Note: (B) Fluorescent, not incandescent. (C) Where first detected.

94. A (pages 424 and 426).

95. B (page 273). Note: Penetrate to the fire and then work back.

96. D (page 167). Note: Use an aerial ladder or ladder tower, not a tower ladder or snorkel, for removing many firefighters from a roof or other area.

97. A (pages 240-241). Note: The first is to have an alternate escape route. The wind is second.

98. B (page 68). Note: Haul up to the floor or the landing below, then via the stairs to the fire floor.

99. A (pages 174-175). Note: (B) 15 to 20 degrees, not 45. (C) Backed in. (D) Fifteen feet past the near end so that you can drive forward if needed later.

100. C (page 447). Note: Use only essential personnel, and rotate them frequently.

PART V

ANSWER SHEETS

General Principles of Firefighting
Answer Sheet

Name _____   Date _____

Class _____   Instructor _____

1. _____

_____

2. _____

3. _____

4. _____

5. _____

6. _____

7. _____

8. _____

Size-Up
Answer Sheet

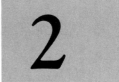

Name _____   Date _____

Class _____   Instructor _____

1. _____
   _____

2. _____
   _____

3. _____
   _____

4. _____

5. _____

6. _____

7. 1. C _____   8. W _____
    2. O _____   9. E _____
    3. A _____   10. A _____
    4. L _____   11. L _____
    5. W _____   12. T _____
    6. A _____   13. H _____
    7. S _____

8. (1) _____ , (2) _____ , (3) _____

9. _____

10. _____
    _____

11. (1) _____
    (2) _____
    (3) _____
    (4) _____
    (5) _____

12. _____

13. _____

14. (1) _____ (most collapse resistant)
    (2) _____
    (3) _____
    (4) _____
    (5) _____ (least collapse resistant)

15. (1) _____
    (2) _____
    (3) _____

16. _____

17. (1) _____
    (2) _____
    (3) _____

18. (1) _____
    (2) _____
    (3) _____

19. (1) _____
    (2) _____
    (3) _____
    (4) _____

20. (1) _____
    (2) _____
    (3) _____

21. _____

Engine Company Operations
Answer Sheet

3

Name _____   Date _____

Class _____   Instructor _____

1. _____

2. (1) _____
   (2) _____
   (3) _____

3. _____
   _____

4. _____

5. _____

6. _____

7. _____

8. _____

9. _____

10. _____

11. _____
    _____

12. (1) _____
    (2) _____
    (3) _____

13. (1) _____
    (2) _____
    (3) _____
    (4) _____

14. _____

15. _____

Hoseline Selection, Stretching, and Placement—Answer Sheet

4

Name _____   Date _____

Class _____   Instructor _____

1. (1) _____
   (2) _____
   (3) _____
   (4) _____

2. _____

3. _____

4. _____

5. _____

6. _____

7. _____

8. _____

9. _____

10. _____

11. _____

12. _____

13. _____

14. _____

15. _____

16. _____

17. _____

Water Supply
Answer Sheet

Name _____   Date _____

Class _____   Instructor _____

1. _____

2. _____

3. _____

4. _____

5. _____

6. _____

7. _____

8. _____

9. _____

10. _____

11. _____

12. _____

13. _____

14. _____

15. _____

16. _____

Standpipe and Sprinkler Operations
Answer Sheet

Name _____   Date _____

Class _____   Instructor _____

1. _____

2. _____

3. _____

4. (1) _____

   (2) _____

5. _____

6. _____

7. _____

8. _____

9. _____

10. _____

11. _____

12. _____

13. _____

14. _____

15. _____

16. _____

17. _____

18. (1) _____

   (2) _____

   (3) _____

   (4) _____

   (5) _____

19. _____

20. _____

21. _____

22. _____

23. _____

24. _____

25. _____

26. _____

27. _____

28. _____

29. _____
_____

30. _____

31. _____

32. _____
_____

33. _____
_____

34. _____

35. (1) _____

(2) _____

(3) _____

Ladder Company Operations
Answer Sheet

7

Name _____ Date _____

Class _____ Instructor _____

1. L _____
   O _____
   V _____
   E _____
   R _____
   S _____
   U _____

2. _____

3. (1) _____
   (2) _____
   (3) _____

4. _____

5. _____

6. _____

7. _____

8. _____

9. _____

10. _____

11. _____

12. _____

13. _____

14. _____

15. _____

16. _____

17. _____

18. _____

19. _____

20. _____

21. _____

22. _____

23. _____

24. _____

25. _____

26. _____

27. (1) _____

    (2) _____

    (3) _____

    (4) _____

    (5) _____

28. _____

29. _____

30. _____

31. _____

32. _____

33. _____

34. _____

35. (1) _____

    (2) _____

    (3) _____

    (4) _____

Forcible Entry
Answer Sheet

Name _____   Date _____

Class _____   Instructor _____

1. _____
2. _____
3. _____
4. _____
5. _____
6. _____
7. _____
8. _____
9. _____
10. _____
11. _____
12. _____
13. _____
14. _____
15. _____
16. _____
17. _____
18. _____
19. _____
20. _____

21. _____

22. _____

23. _____

24. _____

25. _____

26. _____

27. _____

28. _____

Ventilation
Answer Sheet

Name _____   Date _____

Class _____   Instructor _____

1. _____

2. _____

3. _____

4. _____

5. _____

6. _____

7. _____

8. _____

9. _____

10. _____

11. _____

12. _____

13. _____

14. _____

15. _____

16. _____

17. _____

18. _____

19. _____

20. _____

21. _____

22. Diagram quick cut here:

23. (1) _____

     (2) _____

     (3) _____

     (4) _____

     (5) _____

     (6) _____

     (7) _____

24. _____

25. _____

26. _____

     _____

27. _____

28. _____

     _____

29. _____

30. _____

31. _____

32. _____

33. _____

34. _____

35. _____

     _____

36. _____

     _____

37. _____

38. _____

39. _____

40. _____

     _____

41. _____

42. _____

     _____

43. _____

     _____

44. _____

     _____

45. _____

     _____

46. _____

     _____

47. _____

48. _____

     _____

49. _____

50. _____

51. _____

52. _____

     _____

53. _____

     _____

54. _____

     _____

Search and Rescue
Answer Sheet

Name _____   Date _____

Class _____   Instructor _____

1. _____

2. _____

3. _____

4. _____

5. _____

6. _____

7. _____

8. _____

9. _____

10. _____

11. _____

12. _____

13. _____

14. _____

15. _____

16. _____

17. (1) _____

    (2) _____

    (3) _____

18. _____

Firefighter Survival
Answer Sheet

11

Name _____   Date _____

Class _____   Instructor _____

1. _____
2. _____
3. _____
4. _____
5. _____
6. _____
7. _____
8. _____
9. _____
10. _____
11. _____
12. _____
13. _____
14. _____
15. _____
16. _____
17. _____
18. _____
19. _____
20. _____
21. _____
22. _____

23. _____

24. _____

25. _____

26. _____

27. _____

28. _____

29. _____

30. _____

31. _____

32. _____

33. _____

34._____

Private Dwellings
Answer Sheet

Name _____   Date _____

Class _____   Instructor _____

1. _____

2. _____

3. _____

4. _____

5. _____

6. _____

7. _____

8. (1) _____

   (2) _____

   (3) _____

   (4) _____

   (5) _____

9. _____

10. _____

11. _____

12. _____

13. _____

Multiple Dwellings
Answer Sheet

Name _____   Date _____

Class _____   Instructor _____

1. _____

2. _____

3. _____

4. _____

5. _____

6. _____

7. _____

8. _____

9. _____

10. _____

11. _____

12. _____

13. _____

14. _____

15. _____

16. _____

17. _____

18. _____

19. _____

20. _____

21. _____

22. _____

23. _____

24. _____

25. _____

26. _____

27. (1) _____

    (2) _____

    (3) _____

28. _____

29. _____

30. _____

31. _____

32. _____

33. _____

Taxpayers
Answer Sheet

14

Name _____   Date _____

Class _____   Instructor _____

1. _____

2. _____

3. _____

4. _____

5. _____

6. _____

7. _____

8. _____

9. _____

10. _____

11. _____

12. _____

13. _____

14. _____

15. _____

16. _____

17. _____

18. _____

19. _____

20. _____

21. _____

22. _____

23. _____

24. _____

25. _____

26. _____

High-Rise Office Buildings
Answer Sheet

Name _____   Date _____

Class _____   Instructor _____

1. _____

2. _____

3. _____

4. _____

5. _____

6. _____

7. _____

8. _____

9. _____

10. _____

_____

11. _____

12. _____

13. _____

14. _____

15. _____

16. _____

17. _____

18. _____

19. _____

_____

20. _____

21. _____

22. _____

23. _____

24. _____

_____

25. _____

_____

26. _____

27. _____

28. _____

29. _____

30. _____

31. (1)_____

(2)_____

(3)_____

(4)_____

32. _____

33. _____

34. _____

_____

35. _____

36. _____

_____

37._____

Buildings Under Construction, Renovation, Demolition Answer Sheet

16

Name _____   Date _____

Class _____   Instructor _____

1. _____

2. _____

3. _____

4. _____

5. _____

6. _____

7. _____

8. _____

9. _____

_____

10. _____

_____

11. _____

12. _____

13. _____

_____

Fire-Related Emergencies: Incinerators, Oil Burners, and Gas Leaks—Answer Sheet

Name _____   Date _____

Class _____   Instructor _____

1. _____
2. _____
3. _____
4. _____
5. _____
6. _____
7. _____
8. _____
9. (1)_____
   (2)_____
10. _____
11. _____
12. _____
13. _____
14. _____
15. _____
16. _____
17. _____
18. _____
19. _____
20. _____

21. _____

22. _____

23. _____

24. _____

25. _____

26. _____

_____

27. _____

28. _____

_____

29. _____

30. _____

_____

Collapse
Answer Sheet

Name _____   Date _____

Class _____   Instructor _____

1. _____

2._____
_____
_____

3._____
_____
_____

4. (1) _____
   (2) _____
   (3) _____

5._____
_____
_____

6._____
_____
_____

7._____
_____
_____

8. (1) _____
   (2) _____
   (3) _____
   (4) _____
   (5) _____
   (6) _____

9. (1) _____
   (2) _____
   (3) _____
   (4) _____
   (5) _____
   (6) _____
   (7) _____
   (8) _____

10._____
_____
_____

11._____
_____
_____

12._____
_____
_____

13._____
_____
_____

14._____
_____
_____

15. _____

16. (1) _____
    (2) _____
    (3) _____
    (4) _____
    (5) _____

17. (1) _____
    (2) _____
    (3) _____
    (4) _____
    (5) _____
    (6) _____

18. _____
    _____
    _____

19. _____

20. _____

21. _____

22. _____
    _____
    _____

23. _____
    _____
    _____

24. _____
    _____
    _____

25. _____
    _____
    _____

26. _____
    _____
    _____

27. _____
    _____
    _____

28. _____
    _____
    _____

29. (1) _____
    (2) _____
    (3) _____
    (4) _____
    (5) _____
    (6) _____

Final Examination
Answer Sheet

Name _____   Date _____

Class _____   Instructor _____

1. _____	21. _____	41. _____	61. _____	81. _____
2. _____	22. _____	42. _____	62. _____	82. _____
3. _____	23. _____	43. _____	63. _____	83. _____
4. _____	24. _____	44. _____	64. _____	84. _____
5. _____	25. _____	45. _____	65. _____	85. _____
6. _____	26. _____	46. _____	66. _____	86. _____
7. _____	27. _____	47. _____	67. _____	87. _____
8. _____	28. _____	48. _____	68. _____	88. _____
9. _____	29. _____	49. _____	69. _____	89. _____
10. _____	30. _____	50. _____	70. _____	90. _____
11. _____	31. _____	51. _____	71. _____	91. _____
12. _____	32. _____	52. _____	72. _____	92. _____
13. _____	33. _____	53. _____	73. _____	93. _____
14. _____	34. _____	54 _____	74. _____	94. _____
15. _____	35. _____	55. _____	75. _____	95. _____
16. _____	36. _____	56. _____	76. _____	96. _____
17. _____	37. _____	57. _____	77. _____	97. _____
18. _____	38. _____	58. _____	78. _____	98. _____
19. _____	39. _____	59. _____	79. _____	99. _____
20. _____	40. _____	60. _____	80. _____	100. _____